Managing Environmental Crises

Peter Bulleid

Business and the environment – practitioner series
Series editor: Ruth Hillary
Centre for Environmental Technology
Imperial College of Science and Technology

STANLEY THORNES

First published in 1996 by:
Stanley Thornes (Publishers) Ltd
Ellenborough House
Wellington Street
CHELTENHAM
Glos. GL50 1YW
United Kingdom

96 97 98 99 00 / 10 9 8 7 6 5 4 3 2 1

A catalogue record for this book is available from The British Library.

ISBN 0 7487 2439 7

Printed and bound in Great Britain by CIP, Basildon, Essex

TITLES AVAILABLE IN THIS AND OTHER SERIES

On this and the following page is a full list of titles in our TQM Practitioner Series, Business and the Environment Practitioner Series and our new Business Performance Improvement Practitioner Series. Some reviews of individual titles are included below.

> '...one of the best (reports) yet encountered.'
> *Automotive Engineer*
>
> '...a welcome contribution with lots of useful and clearly written information and guidance.'
> *Integrated Environmental Management*
>
> '...clearly written...useful for anyone in business.'
> *Waste and Water Treatment*

Details of where to get further information can be found overleaf.

Title	ISBN	Price (£)
TQM Practitioner Series		
Writing a Quality Manual for BS EN ISO 9000: 1994	0 7487 2421 4	35.00
TQM – The Hands-On Experience: People Make Quality	0 7487 1845 1	35.00
Forms and Other Documents for Use in BS EN ISO 9000 Quality Systems	0 7487 1846 X	35.00
Essential TQM Tools and Techniques: A Practical Business Handbook	0 7487 1847 8	35.00
Gaining and Benefiting from ISO 9000 Registration: A Step-by-step Guide (Second Edition)	0 7487 1837 0	35.00
A Quality Auditing Manual (Second Edition)	0 7487 1840 0	35.00
Sustaining Quality Advantage in Financial Services	0 7487 1828 1	35.00
Beyond the BS 5750/ISO 9000 Certificate: The Bureaucracy Buster's Guide to Quality Assurance	0 7487 2106 1	35.00
Writing an ISO 9000 Quality Manual: Practical Guidelines	0 946655 89 8	35.00
Risk Management in a TQM Environment	0 7487 1905 9	35.00
FMEAs: A Manager's Handbook	0 7487 1861 3	35.00
TQM: Its Impact on Bottom Line Results	0 946655 96 0	35.00
The Role of Junior and Middle Level Management in TQM	0 946655 74 X	35.00
Quality Function Deployment	0 946655 72 3	35.00
The Role of the Modern Quality Manager	0 7487 1867 2	35.00
Measuring Service Quality	0 946655 75 8	35.00
TQM in Sales and Marketing	0 946655 76 6	35.00
TQM: Does It Always Work?	0 7487 1936 9	35.00
Communication in Successful TQM	0 7487 1872 9	35.00
Kaizen	0 7487 1901 6	35.00
TQM Training	0 7487 1871 0	35.00
Teams and TQM	0 7487 1870 2	35.00
QA and ISO 9000/BS 5750 in the Hospitality and Catering Industries	0 7487 1869 9	35.00
BS 7750: What the New Environmental Management Standard Means for your Business	0 946655 60 X	35.00
TQM in Service Industries: A Practitioner's Manual	0 7487 1914 8	35.00
Managing and Improving Service Quality and Delivery	0 7487 1907 5	35.00
Competitive Benchmarking: An Executive Guide	0 7487 1926 1	35.00
TQM-Based Performance Measurement: Practical Guidelines	0 946655 52 9	35.00
Quality Costs: Their Impact on Company Strategy and Profitability	0 7487 1875 3	35.00
Customer Care: Strategy for the 90s	0 7487 1868 0	35.00
Ensuring Your Business Achieves and Profits from BS 5750 Registration: A Practical Guide	0 946655 44 8	35.00
Leadership of Customer-Driven TQM	0 7487 1865 6	35.00
TQM in the Electronics Industry: A User's and Supplier's Guide	0 7487 1874 5	35.00
Software Quality Assurance: What It Buys You and What It Costs You	0 7487 1902 4	35.00
Implementing TQM in Small and Medium-Sized Companies	0 946655 59 6	35.00
Selling Professional Services	0 7487 1873 7	35.00

Continued overleaf

Business and the Environment – Practitioner Series

Managing Environmental Crises	0 7487 2439 7	35.00
The Role of the Environmental Manager	0 7487 1842 7	35.00
Environmental Compliance Auditing	0 7487 2187 8	35.00
Measuring Environmental Business Performance: A Comprehensive Guide	0 7487 2063 4	35.00
Environmental Improvement Through the Management of Waste	0 7487 2129 0	35.00
The 1994 Waste Management Licensing Regulations and the New Definition of Waste	0 7487 2133 9	35.00
International Environmental Policy and Management	0 7487 2132 0	35.00
Integrated Pollution Control	0 7487 2061 8	35.00
Life Cycle Assessment	0 7487 2131 2	35.00
The Implications of Water Regulation for Industry	0 7487 2062 6	35.00
A Trouble Shooter's Guide to Environmental Management Systems: BS 7750: 1994 and Other Options	0 7487 1841 9	35.00
The Eco-Management and Audit Scheme: A Practical Guide	0 7487 1900 8	35.00
Business Success from Seizing the Environmental Initiative	0 946655 93 6	35.00
Improving Environmental Performance: A Proven and Effective Approach	0 7487 1864 8	35.00
Managing Environmental Risks and Liabilities	0 7487 1906 7	.35.00
Contaminated Land: Assessment and Redevelopment	0 7487 1937 7	35.00
The Environment and the Planning System: Business Implications	0 7487 1866 4	35.00
Business and Environmental Accountability: An Overview and Guide to the Literature	0 7487 1863 X	35.00
Environmental Auditing: A Guide to Best Practice in the UK and Europe	0 7487 1862 1	35.00
Directors in the Dock	0 7487 1927 X	35.00

Business Performance Improvement – Practitioner Series

Getting the Right Product	0 7487 2453 2	35.00
Writing Effective Performance-Related Procedures	0 7487 2503 2	35.00
Corporate Communication and Business Performance	0 7487 2572 5	35.00
Quality In Business Performance: How to Change Attitudes	0 7487 2438 9	35.00
Training Industrial Staff: A Practical Handbook	0 7487 2369 2	35.00
Communicating for Improved Business Performance	0 7487 2266 1	35.00
Getting Closer To Your Suppliers	0 7487 1852 4	35.00
Customer Satisfaction Measurement: A Practical User's Guide	0 7487 2138 X	35.00
Competitive Customer Service	0 7487 2135 5	35.00
Getting Good Value from Consultants	0 7487 1851 6	35.00
Commercial Awareness: Increasing Your Edge	0 7487 1838 9	35.00
Business Control, Accountability and Corporate Governance	0 7487 2134 7	35.00
An Implementation Guide on How to Re-engineer Your Business	0 7487 2130 4	35.00
Practical Implementation of New Management Strategies: Managing Your Processes	0 7487 1839 7	35.00
Ensuring Your Business Survives: A Self Analysis and Self Help Manual for Managers	0 7487 1844 3	35.00
Self Assessment: A Tool for Integrated Management	0 7487 2060 X	35.00
Better Planning for Performance Improvement	0 7487 1853 2	35.00
Changing the Business: A Practical Guide to Performance Improvement	0 7487 1822 2	35.00
Business Process Re-engineering: A Practical Handbook for Executives	0 7487 1821 4	35.00

If you would like further information, please contact us at the address below. Many companies buy bulk copies of our reports; if you are interested in doing this, we would be pleased to discuss discount levels.

Customer Services Department
Stanley Thornes (Publishers) Limited
Ellenborough House
Wellington Street
CHELTENHAM
Glos. GL50 1YW

Telephone: (01242) 228586
Fax: (01242) 221914

CONTENTS

Part 1 INTRODUCTION

1. What is a Crisis?

This is not such an obvious question as it may first appear. A house fire in Huddersfield is hardly going to constitute a crisis to most readers. To the reader whose home it is, the fire is most certainly a crisis. Clearly therefore geography is an element in its definition. The closer to home the incident is, the more likely it is to be a crisis.

However, consider now that the fire is not in a dwelling, but in a nuclear reprocessing plant and that it results in the release of plutonium into the atmosphere. Immediately we are faced with a crisis that is not just local but regional, if not global. The experience of Welsh sheep farmers, who suffered restrictions on their business activities following the fallout from Chernobyl, will testify to that. Thus, it is not so much the point of release as the location of the impact, that is the geographical factor.

Unfortunately there are more factors to the equation. Plutonium for example is both radio-active and a toxin in its own right. The total spectrum of the effects that arise from an uncontrolled release can be both diverse and time dependant. The crisis might appear to die down fairly quickly as the media lose interest. Nevertheless the environmental effects may continue to be felt for many years, perhaps for generations and they may change over time. When, therefore, does the crisis end and how can time be built into the definition? For that matter when does it begin if it is, for example, a slowly accumulating ecotoxin?

Mention of the media brings another factor into the equation and one with which we will be dealing in more detail in Chapter 9, namely that of public perception. What may appear to be a straightforward operational issue can be turned into a crisis for the company by other parties, as Shell found out with Brent Spar. Likewise the shere complexity of the environment itself can make crisis identification and management more of an art than a science. As our scientific knowledge increases, then activities that were formerly taken for granted may come to be regarded as dangerous and their effects give rise to a crisis almost overnight.

Curiously the opposite can also happen. The loss of the oil tanker Brear and her cargo of oil in the Orkneys appeared to be both an immediate and long term crisis. However, because of the nature of the oil, and the weather conditions at the time of the spill, the effects were relatively local and short lived. Quite what happened when the oil dispersed and what the possible longer term effects are, in many cases is still unknown. The crisis, at least in the eyes of the public, soon abated of its own accord.

The E.C. "Seveso" Directive attempts to take all these factors into account when it defines an environmental crisis as:

"permanent or long-term damage to particular unique, rare or otherwise valued components of the man made or natural environment or if there is widespread environmental loss or damage."

The problem with this definition is that it is imprecise and potentially very wide ranging. Also it is written from a legislator's point of view rather than that of a business enterprise.

It is clear that the definition of what constitutes a crisis is extremely difficult unless we restrict our view considerably. Accordingly we shall define an environmental crisis as **"any uncontrolled or unplanned interaction between the business and the environment that could result in significant costs."** These costs can arise from a number of sources such as:

- disruption to or loss of production
- fines or legal costs
- clean up costs
- restrictions placed on business activities by regulators
- unfavourable publicity
- increased insurance premiums.

This list is far from comprehensive. The use of this definition avoids the problems of trying to define issues such as magnitude, timescale, who the recipients of the effects are and some of the other difficulties briefly explored above. However, it must always be borne in mind that adopting this relatively simplistic and limited definition will not necessarily place the firm in the same position as others who have an interest in the environment. To a regulatory body, the issue of cost to the business may be a minor consideration, while to the public it may not be a consideration at all. In this context it should be recognised that placing a value on the environment is extremely difficult. We shall explore this further in later chapters but the issue of "commons", that is aspects of the environment that are shared by all such as air, water and scenic beauty, are liable to always give rise to dispute. Legislative valuation as evidenced in fines or clean up costs or both, are seldom likely to reflect the value that society places on commons, so that there is often a disparity between the perception of magnitude, as measured by crude monetary values, between business and the public.

Brent Spar - Doing the Right Thing the Wrong Way

In July 1995, Shell UK, part of one of the largest multi national companies in the world, with considerable resources and experience in the environmental field, found itself in an environmental crisis and more seriously, lost control of it. By 1992 the Spar mooring and storage facility in the Brent oilfield, was time expired and surplus to requirements. A plan for its disposal was formulated that met current requirements, after careful consideration of the available options, and was agreed with all relevant authorities. The case for disposal of the Brent Spar at sea was put in Shell's press briefing of February 1995 which is reproduced in Appendix F.

However, the plan was opposed by some environmentalists, notably Greenpeace. The group clearly disputed Shell's figures but, more importantly, it opposed the principle of sea disposal. Its case was put in a press briefing dated April 1995. It seems likely that, with hindsight, the balance of informed opinion will show that sea disposal was the best practicable environmental option. Nevertheless, Greenpeace was able to mobilise public opinion and Shell lost the battle for hearts and minds. It is likely to cost them many millions of pounds as a result. How could it be that in doing the right thing the company found that it all went wrong?

There is no easy answer to the question. A number of issues are explored below all of which have a bearing on the matter to a greater or lesser extent. No doubt others will emerge. It may be that time will show Shell to have been unlucky and a victim of circumstance but all of the points discussed could have relevance to the management of future cases by others. They are worth bearing in mind when considering the various strands to the crisis management strategy advanced in this book.

* **Life cycle planning.** The Brent Spar was conceived and designed over 20 years ago and little thought would have been given to its eventual decommissioning and disposal. In the heady rush to get North Sea oil on stream at a time when environmental knowledge and legislation was much more basic, such issues seemed remote. Today they are less so. A high quality environmental management programme for a new project under BS 7750 would now encompass this aspect. The Construction, Design and Management Regulations require a file to be created and updated over the life of a project to facilitate decommissioning. The EC is promoting life-cycle analysis for various categories of products. In other words the climate has changed considerably already and will continue to do so.

  Whether an earlier consideration of decommissioning and disposal of the Spar would have avoided the problem is doubtful. However, it would have given more time for scientific and technical studies, gaining a consensus view and planning to avoid the "emergency", than was the case. It may be that some of the problems such as contaminated sludge or radioactive scale could have been foreseen and avoided. It would certainly have been both precautionary and good planning, emergency or otherwise, to have considered disposal at the outset and to have identified the possible results including intense public interest.

  This is not to say that such foresight is easy. While plans can be constantly viewed and updated, there is clearly going to be a considerable element of uncertainty even when looking only two decades ahead. What realistically could the railways have done, for example, when decommissioning and abandonment of some facilities lay 150 years in the future?

* **Consultation.** Shell had consulted all the relevant statutory authorities and commercial interests. It had also gained the support of the British government and at least the acquiescence of several foreign governments. Yet this was not enough. It had not consulted all the pressure groups that were of sufficient size and organisation to mount a campaign against the proposal. Again such consultation may not have avoided the problem but it would at least have forewarned Shell of the likely opposition, allowing counter moves to be planned.

  There is a natural tendency to fight shy of consulting possible opponents on the basis that it is unlikely to change their minds and it will only be giving them ammunition to fire back. However, like trying to keep the press at bay, a no consultation strategy is likely to prove counter productive.

* **Government Support.** Shell believed, not unreasonably, that having the support of the government provided them with a powerful ally. Indeed the agreement of the government was a pre-requisite from a regulatory point of view, but agreement and support are two different things.

  Many environmental pressure groups are at loggerheads with the government. Policy, priorities and the relatively slow rate of change are just some of the bones of contention. Opposition to the government can verge on paranoia and a willingness to believe in conspiracies. Thus government support can become, in some people's minds, an indication that there is something to hide. Both big business and government ministers being perceived as closing ranks to push through the easy option against the best interests of the environment.

This is a particularly difficult one for a company to win. It is likely to be damned if it has government support and damned if it does not. Perhaps the best approach is openness and, as mentioned above, early and thorough consultation. This, coupled with avoiding being to closely aligned with government policy and endorsement, is a promising strategy.

* **Track Record.** Shell has put much time and effort into the environment and can show many solid achievements. Yet it is not perfect. Given the scale of its operations it would be astonishing if it were. The ENDS Report for June 1995 noted Shell's difficulties over Brent Spar and went on to report that in addition the company was:

    - Appearing in court in Lancashire charged with polluting an aquifer
    - Facing prosecution over groundwater on Merseyside
    - Convicted of breaching a discharge consent in Carlisle with fines and costs of over £30,000 while readers were reminded of a £1 million fine it received in 1990 for pollution. (Note how poor past performance can continue to haunt a company for many years. Its continuing to do so is here in this book!)

  Regardless of the merits of the Brent Spar proposal, it is obvious that Shell has a problem with winning hearts and minds on environmental issues with this sort of performance. It is possible for quite minor misdemeanours to have an adverse effect on major projects out of all proportion to their size. The point must be made that good emergency planning needs to be part of good environmental planning in general.

  Ultimately scientific argument and government support could not counter moral and commercial pressure and, as we all know, Shell changed its plans and is now exploring other disposal options for the Spar. Greenpeace may have difficult decisions to make in being party to an option that is perhaps environmentally less than the best. Shell released an explanation for its actions on July 10th 1995 and one must have some sympathy for its tone of bewilderment and the questions posed in the second but last paragraph.

  From this example we see that there may be a large number of "stakeholders" in the environment, and therefore in an environmental crisis, and what is more you do not actually have to damage anything to have a crisis. Besides the company itself there may be shareholders, investors and customers, while on the regulatory side there are a range of authorities, government and non-governmental agencies and the emergency services. In addition the "public" includes a range of groups, from the international to purely local or single interest and then there is also the media. All these parties are likely to have differing views on what constitutes a crisis and what should be done about one. It is most unlikely that there will be any unanimity of views and agreed "right course of action". It will be readily appreciated that this greatly increases management's problems. In satisfying one stakeholder it is likely that another one will be dissatisfied.

This book is intended to provide managers unfamiliar with the environment and emergency planning with the basic knowledge and tools to deal with an environmental crisis. To do this we need to understand some basic environmental mechanisms, what may cause a crisis, and the legislative framework in which business must operate. We will then explore crisis avoidance and management techniques.

What is the Environment?

Having defined what constitutes a crisis it is worth looking at the factor that is central to this book, namely the environment. It would be pointless to try and explore the nature of the environment here; the subject is too vast. Readers with wider interests in the environment and who wish to explore it further are referred to the bibliography at the end of the book where a few key publications are identified. All we can aim to do is to provide managers with an overview of some of the principal environmental mechanisms.

The first issue should be to define what we mean by the environment. There are many possible definitions, but for the purposes of crisis management it is necessary to cast the net wide so as to avoid the possibility of missing potential problems that could develop into crises. In this respect perhaps the best definition is that the environment is "everything outside the business itself." Now clearly this definition needs to be applied with some common sense. We are not dealing with, for example, the financial or commercial "environment" within which the business operates. It is the physical and biological environment, and crises that arise from the interaction of the business with it, that concerns us here.

Unfortunately for the manager the physical and biological environment is so extensive, in both geographical and functional terms, as to be a lifetime study in itself. However, if the manager is to be able to manage crises effectively, it is necessary to understand certain basic facts and principles which underlie most environmental systems. Set out below are the four key issues that are likely to affect decisions as to why and how crises should be managed.

Complexity

The environment is a very complex system and, as a result, disturbances in one part can have wide repercussions, the reasons for which are sometimes difficult to understand. It is also unpredictable, that is although we may have some idea at to how different elements interact, it is not possible to be certain about cause and effect chains and to make quantitative predictions with any confidence. One reason for this is because the environment is dynamic. It can change spontaneously without the hand of man and any disturbance caused by man may retard, accelerate or redirect what are underlying and naturally occurring changes. In this context it should be remembered that environmental crises may occur naturally. Volcanic eruptions, floods, fires and other disturbances are frequent occurrences world-wide and in some areas are a vital part of the local ecology. Crises can be caused by ignoring or attempting misguided control of these phenomena. Desertification and the salination of soils caused by over-exploitation of the land or artificial irrigation systems are but manifestations of man's inadvertent creation of environmental crises in attempting to work against natural processes.

Cycles

Many processes are cyclical in nature. Thus, certain elements and substances such as carbon, oxygen, nitrogen and water are constantly caught up in cycles that interact with, and are of vital importance, to life. While some disturbance to these cycles can be tolerated at local level, gross disruption of the natural cycles can be very dangerous and the necessity to maintain and protect them in the long term cannot be over emphasised. Furthermore, many other cycles are found occurring naturally. Seasonal change is obvious, but some animal populations, for example, naturally vary in a cyclical pattern over time. As a result it may not be obvious if a change is natural and self correcting or manmade and irreversible, until it is too late. Cycles may be global in scale so that local, adverse effects appear to have minimal consequence but we do not fully understand the extent to which local disturbances can combine to become global.

Food Webs and Communities

All life relies on the capture of solar energy by plants and its storage as food compounds. Animals are ultimately dependant on plants as a food source, either directly as in the case of herbivores, or indirectly as carnivores or detritivores (generally very small animals that act as nature's cleaners-up of "rubbish"). Energy can be envisaged as flowing through ecological systems, not so much as through a chain, as through a web. This is both a strength and a weakness. Loss of a species in a web may have little effect, as in some cases other species can switch their diet or habits to compensate. On the other hand, the loss of one particular species may adversely affect several other species, causing collapse of the web and severe environmental consequences. Curiously the role of the top or final species in the web can often be crucial. This species is sometimes known as the "keystone" and it is both particularly vulnerable to disturbances and at the same time often vital to the whole functioning of the local ecology. Man is increasingly placing himself in the role of keystone species.

Some species are particularly sensitive to changes in their environment and as a result one of the first indicators of a gradual environmental change may be an alteration in the mix of different species that go to make up the community of animals or plants in the area affected. Certain sensitive species can therefore be used as indicators of environmental health. Traditionally human health has been the yardstick by which risk and damage are assessed. Unfortunately in terms of the environment this is not a very good indicator. Human life can continue in a severely degraded environment, given sufficient resources. Such a situation is neither desirable nor sustainable.

Stability

Notwithstanding natural disturbances, many environmental processes are self regulating and adjust automatically to take account of changes. There is a tendency for environmental systems to be stable and for them to be able to absorb a certain amount of damage. However, this resilience to external effects is limited and once the limit is reached then the system can become unstable, so that change then continues to occur spontaneously and at an increasing rate. The situation can then run out of control. The limits to stability are seldom obvious or necessarily known. Manmade disturbances from different sources may cancel each other out, but are more likely to have a multiplier effect, both on themselves and upon naturally occurring disturbances.

Some changes to the environment are irreversible while others may in theory be reversible, but in practice be irreversible. For example, releasing a genetically modified organism that hybridises with the natural type, or manufacturing and distributing a very stable chemical, are irreversible for all practical purposes. Clearly considerable care needs to be exercised in such cases.

Critical Issues

The critical issues for the environment will vary from location to location and will be influenced by the nature of the proposed or actual changes being brought to the eco-system. However, globally and in the long term, there is agreement that two issues stand out. These are:

- The loss of genetic material through the extinction of species, and
- The loss of non-renewable resources.

It is in these areas that the environment faces its greatest risk. While these are not the same as business risks, commercial development or activities that lead to either of these consequences will not be sustainable in so far as future generations are concerned. They should therefore be avoided as if they were direct business risks.

In summary it is well for the crisis planner or manager to remember:

- The environment is an inherently difficult system to manage.
- There are many potential interactions with it.
- There is a great deal of uncertainty.
- Once something has gone wrong it may be a very laborious and expensive exercise to correct it, if indeed it can ever be corrected.
- Certain environmental assets and processes are critical to all life.
- Increasingly the public are concerned about environmental issues and may be very well informed.

Given these facts, it is clearly going to be in the interest of any business to manage, if not eliminate, potential environmental crises and therefore unquantifiable costs if possible. The purpose of this book is to assist all business managers but particularly those in small and medium sized enterprises to do just that. Such businesses are particularly vulnerable. The effect that they have on the environment, particularly in terms of risk and potentially precipitating a crisis, can be as great as for a large corporation. However, they are unlikely to have either the system or resources to cope with a crisis. Lacking these, not only is a crisis more likely, but the effect it could have on the business will be that much greater.

Set out in the following chapters is guidance and suggestions that will enable companies to deal effectively with crises without recourse to the commitment that would be appropriate to a large multi-national. It is aimed at the environmentally responsible manager with limited resources. However, the principles outlined are equally applicable to larger companies and can either be directly applied by them or used as a benchmark against which their own system can be judged.

Difficult Management Decisions

The case of the Exxon Valdez is a good example of some of the difficulties faced by an industry in an environmental crisis. It is also an extreme example of the costs involved. The oil industry is not a major marine polluter through oil spills and accidental discharges. However it is perceived to be "dirty" by the public no doubt due in part to the poor practice of many ship owners in the past of dumping waste oil from bunkers and tanks at sea.

Although conscious of the need for good operating practices, the avoidance of accidents and rapid response to emergencies, incidents such as the Torrey Canyon and Amoco Codiz, showed that the industry was not well prepared. However the Exxon Valdez dramatically exposed the shortcomings of Exxon itself, the industry and the environmentalists. Two particular lessons stand out.

- There was a crucial delay in the reaction to the grounding. This led to the loss of the initiative both in terms of containment of the oil and in terms of control over the incident, media reporting and public reaction.
- The equipment and techniques then available for containment and clean-up was of limited effectiveness.

As a result the industry now takes much more interest in pre-planning for emergencies and in the design and stockpiling of emergency equipment. There is also much closer co-ordination between the industry and the various authorities such as the coastguard. Later in this book we shall explore in detail how rapid and effective response to incidents can be planned in advance.

There were other interesting lessons to be learned from this incident. The public outcry and dire predictions of "experts" as to the likely damage led to a desire for a total clean-up. As a result, within three days more equipment had arrived at the scene than could be effectively used by the local resources. Here as in similar situations both before and after, too great a reliance was placed on heavy mechanical equipment for beach cleaning. The equipment itself damaged the environment without significantly improving the situation. Low technology, manpower intensive, clean-up methods are often more effective and less risky. However, the wisdom of hand cleaning individual pebbles, as was resorted to in Prince William Sound, must be of dubious value except as a public relations exercise. Hindsight has shown that natural recovery by the eco-system has been faster than predicted and that the intensively cleaned areas have recovered slowest. The point has already been made that the natural environment is an extremely complex system. Having once accidentally damaged it, it may at times be best to leave it well alone rather than rush in with an ill-considered action. Clearly the need for a quick but an effective response can only be fulfilled by pre-planning and better understanding.

Cost emerges as the other significant consideration. The value of the cargo and cost of repair to the ship were about 25 million dollars - a not inconsiderable sum. However, this pales in insignificance against the estimated 2 billion dollars spent on the clean-up and other direct costs to the company. But it does not end there. The oil industry as a result of the Exxon Valdez incident now has to contend with:

- New environmental legislation
- Research cost
- Tanker design and production costs
- Extra insurance
- Additional operating costs
- Damage to reputation
- Corporate restructuring
- Provision of emergency equipment.

Bearing in mind that this list is not exclusive and that the oil spill clean up costs alone are 900 million dollars, the estimated additional cost of in excess of 2 billion dollars is perfectly credible. This is a salutary warning to those who would argue either that we do not need, or that we cannot afford, environmental crisis management.

2. The Possible Causes

The possible causes of potential crises will obviously vary from company to company and from business sector to business sector and it would be impossible to deal with them all in a single book. Fortunately the root causes are common and so are the basic systems for managing both them and any subsequent crises. We can, therefore examine these and leave it to the individual manager to interpret and apply them to his own particular circumstances.

Any process, or any business come to that, can be viewed as a black box. Into that box are put raw materials including energy and out of it comes a product and, to a greater or lesser extent, wastes. Each of these inputs and outputs and the black box itself has the potential for causing an environmental crisis and we will examine each in turn. (see Figure 1)

THE BLACK BOX

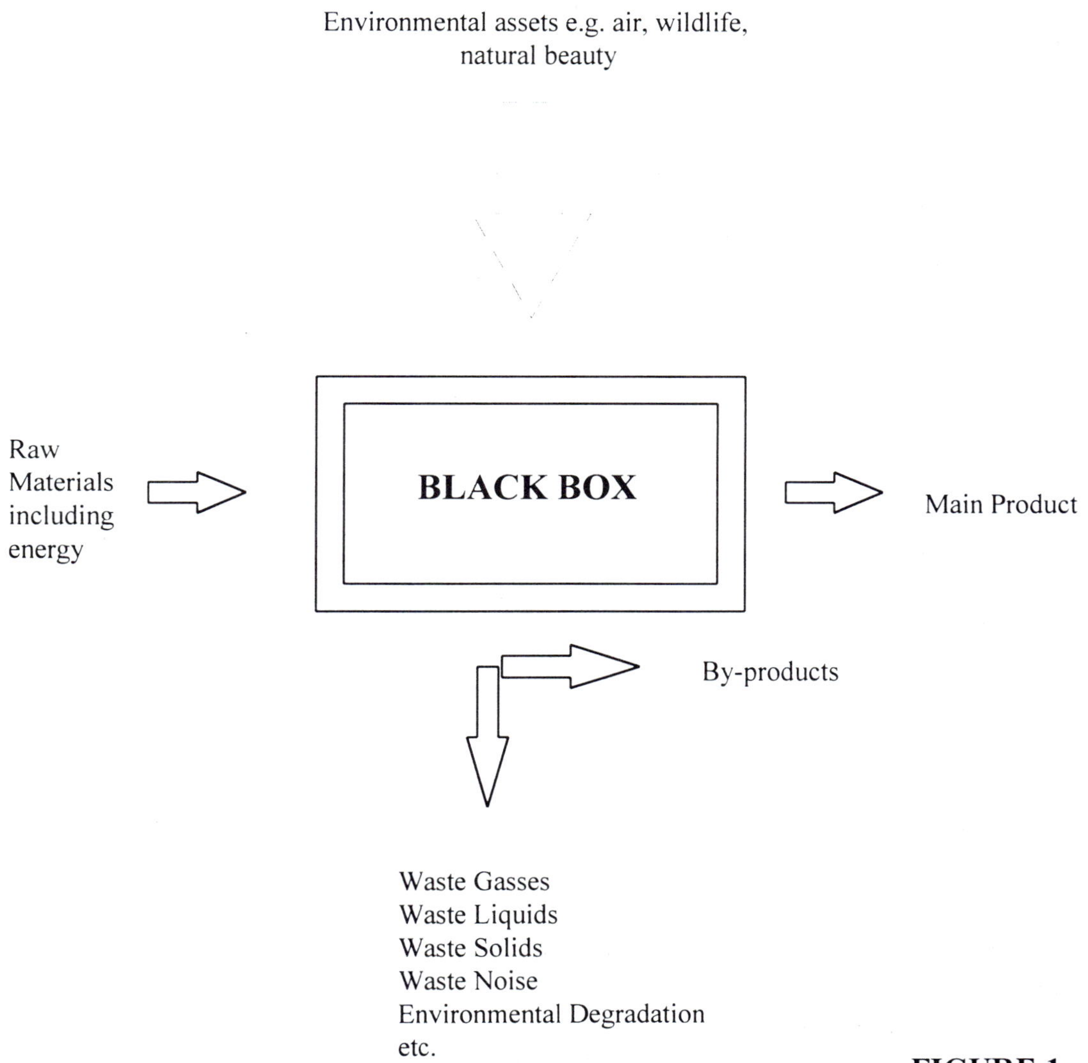

FIGURE 1

Inputs

At first sight the likelihood of an input giving rise to a crisis is small. However, returning to our definition, where crisis is equated to an uncontrolled environmental cost, then two possible causes can be identified.

The first is with the sourcing of raw materials. Ultimately this goes back to natural resources, the depletion of which can cause considerable environmental disturbance, at least on a local scale. Globally, the cumulative effects can be equally severe, as the loss of tropical forests illustrates. Even where the resource is perhaps plentiful and commonly available, such as water, its abstraction may be subject to controls or the "downstream" effects of extraction may be very damaging to the local environment. In any event substantial costs to society may arise over and above the basic cost of the resource. While at present the risks associated with resource utilisation seldom lead to crises, this situation is unlikely to hold in future as demand rises and public expectations and regulatory controls increase.

The second potential cause relates to the control of the quality of the inputs. An industrial process or a service, even one that is essentially intellectual in nature, will be designed to operate with inputs of a particular nature or quality. If the nature is wrong or the quality faulty, then damage can result to the process or product. An explosive release of reactants arising from contamination of a raw material or an incorrect decision based on faulty data can both result in an environmental crisis.

Outputs

These are the intended results of the process and are presumably the marketable product of the business. It would be nice to think that the product will be environmentally benign. Asbestos products, polychlorinated biphenols, tributyl tin, numerous biocides, unfortunately the list is seemingly endless of products that were initially hailed as panaceas but, when faced with the test of time, have proved to be environmental disasters. While the cost to the individual manufacturer may not always have been sufficient to constitute a crisis, the costs to industry as a whole or to society are such as to constitute one. Business in the end faces these costs through increased taxation and regulatory control.

Lest it be thought that the costs to the individual business are always going to be small from this cause then it is worth considering what the cost of a product recall might be, let alone the cost of replacing the product altogether, as has happened more than once with washing powders for example. It is your product that will be in the marketplace so it is vital that the environmental risks and consequences associated with it are assessed and managed.

Wastes

There is no doubt that the greatest potential for causing environmental crises rests with wastes. These can be defined as:

- Controlled and uncontrolled emissions to air
- Controlled an uncontrolled discharges to water
- Solid and other wastes
- Contamination of land
- Noise, odour, dust vibration and visual impact
- Heat
- By-products that are wastes that have some further use.

By their very nature wastes tend not to command attention and to be easily overlooked. However, they also have to be disposed of in the environment, often in a local concentration that may either last a considerable period of time or rapidly escape from any form of control. They are not wanted by anyone and can potentially cause annoyance or damage or both. It needs little imagination to see that they can readily give rise to additional costs to the business of a magnitude and suddenness that can constitute a crisis. For this reason much of the current environmental legislation is directly or indirectly concerned with the management of wastes.

The other fact to bear in mind about wastes is that they have considerable potential for causing conflict with neighbours. For example complaints from local residents has resulted in Her Majesty's Inspectorate of Pollution serving a notice on Coalite Smokeless Fuels at Bolsover over the issue of dust spreading to neighbouring properties. This is rather ironic in view of the nature of the product. In fact many of the complaints to the environmental health officers of local authorities concern odours or noise. These are wastes as much as solids and liquids and can be most intractable to mitigation or control and hence costly to the company.

As a final thought on the subject, remember that waste is waste. That is to say that it is in this area that significant cost savings can usually be found; perhaps enough to fund the setting up of environmental or crisis management systems. Savings can be made on both the cost of materials and their handling. They can also be made on disposal costs, an increasingly important area in view of the new landfill tax. Digital is one of many companies that has been looking closely at waste. It has achieved a 5,000 tonne reduction in waste and packaging over three years. As an additional benefit, reduced waste leads to reduced business risk from adverse environmental effects.

The Black Box

The final element in the equation is the black box itself. The design, operation and maintenance of the box, whether it be a process or a plant, are all potential causes of crises. Costs can arise from excessive consumption of resources, leaks, unwanted by-products or poor quality control, all of which can translate into an environmental crisis. The black box must be considered, not just from the point of view of normal operating conditions, but also from that of abnormal ones, including shut down and start up conditions, incidents and accidents or non environmental emergency situations.

As an example of the problems that can occur with waste output from the black box the case of Miller Group of Edinburgh is instructive. In late1994 it was found guilty on four charges of discharging surface water contaminated with silt into two watercourses and fined £40,000. It should be noted that silt can smother fish and invertebrates and also reduce plant photosynthesis and that it can have a very detrimental effect on aquatic ecology.

The black box in this case was the construction of a road, which might not at first sight appear to be an industrial process. Nevertheless it does take in raw materials and provide both a product and wastes. Nor might removing excess rainwater to a watercourse in a particularly wet period, appeared to be anything untoward. Unfortunately, abnormal operating conditions resulted in a substantial fine through the neglect of adequate control. Sanctions are there that can turn what might appear to be an inconvenience into a crisis.

I must hasten to say here, that a number of examples are given throughout this book of companies that have failed to avert an environmental crisis. I should make it clear that in citing them I do not mean to imply criticism of the individual company. They are used only as examples of the problems that can beset companies in general and no doubt, some at least, have learnt a painful lesson and are now implementing management systems to control the risks and improve their environmental performance.

Management and Risk.

This introduces two key elements that build together to create a crisis management system and which recur throughout this book, namely management and risk. Later chapters deal with both in detail, but both form part of a more extensive system termed environmental management. This book does not attempt to deal with management systems as a whole as the subject is too great. Readers are advised to turn to "A Trouble Shooters Guide to Environmental Management Systems: BS 7750: 1994 and Other Options" in this series for a full exposition of systems currently being implemented. Nevertheless, just as we have had to consider some fundamental concepts of ecology, it is necessary to consider the key elements of an environmental management system to see how crisis management fits within it.

Environmental management is not some arcane science unrelated to any other activity that the management of an enterprise might normally expect to undertake. It can be and it should be, closely integrated with general line management. It can also share many elements in common with the management of issues such as health, safety and quality. This much is recognised in the main instrument in use in this country - BS 7750 Environmental Management Systems. While this is by no means the only model, the forthcoming international environment system ISO 14001 and the European Unions Eco-Management and Audit System (EMAS) both have similar functions, it is the most completely developed and widely used one, being adopted for use in some other countries besides Britain. For this reason we shall briefly consider its main elements, which are set out below:

1. Ensure there is management commitment to the environment and adopt an appropriate policy.
2. Audit and record the effects that the company currently has on the environment.
3. Set objectives and targets for the purpose of achieving the policy.
4. Identify who will be responsible for doing what.
5. Provide the procedures and resources to enable them to do their tasks.
6. Set up auditing, corrective and change control mechanisms.
7. Regularly review performance and adjust the system accordingly to ensure that there is continual improvement.

While criticisms can be levelled at the Standard, such a system has a number of things to recommend it. Firstly, it is systematic and forces management to confront environmental issues. Secondly, it is not prescriptive in terms of performance, beyond the matter of legislative compliance, so allowing the firm to set its own targets and proceed at its own pace. Finally, it embodies the concept of continual improvement so that from modest beginnings performance will improve over time. Any company that has significant effects on the environment or which operates in a market that is sensitive to environmental matters is strongly advised to consider implementing such a system if it has not already done so.

In so far as crisis management is concerned it is notable that the BS 7750 has some specific requirements. We have already seen that the outputs from the black box need to be considered in both abnormal and emergency situations. This requirement is explicitly stated in connection with setting up a "register of the environmental effects." It is, however, in connection with the production of a manual setting out the main elements of the system and also the operating procedures that the Standard states:

"In addition with dealing with the normal activities of the organisation, the manual shall cover abnormal operating conditions, and incidents, accidents and potential emergency situations."

As an aside, to show that a crisis can in fact occur during non operational situations, Medway Power received an HMIP enforcement notice during the commissioning of its Isle of Grain power station. Contractors fractured a pipe for dosing cooling water with biocide and released sodium hypochlorite into the Medway. HMIP decided that the problem was operator error, since no action was taken to control the leak, and required a retraining programme to be implemented before operation commenced.

The British Standard then goes on to specifically cover the provision of plans to deal with crisis situations:

"Emergency plans shall, where necessary, contain relevant environmental information and instructions. The emergency plans shall be tested, wherever practicable, by methods capable of checking their suitability and effectiveness."

The Standard is therefore quite specific in its requirements for devising and practising, if possible, measures for crisis management. It also states in the commentary on the various elements of the system:

"The site emergency plan and occupational health and safety manuals, or documents, should also incorporate relevant environmental information and instructions."

so seeking to tie together different elements of emergency planning and cross linking with such matters as the Control of Industrial Major Accident Hazards (CIMAH) Regulations, which we consider in more detail in Chapter 8.

Although not mentioned specifically by the standard, risk is a key factor in emergency planning. Chapter 5 deals in some detail with risk in relation to environmental hazards and ties it in with operations and other matters. We will be looking in a bit more detail at the operational side of management systems in Chapter 6.

The Four Lines of Defence.

It is now possible to define the various measures that are to be used to manage any environmental crisis. (See Figure 2). Initial concerns centre around risk. Unfortunately as we

have seen, the environment is an exceedingly complex system so that normal risk management is not always applicable. As a result the first line of defence has to be the precautionary principle (Chapter 4). Conventional risk analysis and management then becomes the second line (Chapter 5).

The concept of a comprehensive management system, which we have introduced above, gives the third line, namely the operational control of activities. Finally the fourth line of defence is the emergency plan itself. The first three lines are directed at ensuring that emergencies do not arise or, if they do, then their likely nature, frequency, location and magnitude have been identified. The final defence is dealing with, containing and minimising a crisis that has occurred in spite of the other measures.

As a crisis will most probably involve other parties than just the company itself, for example the emergency services or various regulatory bodies, then Chapter 8 looks at how these parties plan for and deal with environmental emergencies. As a crisis is most unlikely to pass by without the press or public becoming aware of it, Chapter 9 considers the public relations aspects and what is termed reputation risk management.

However, before exploring these lines of defence, it is advisable to consider the regulatory and legal framework within which both day to day operations and crisis situations must be managed.

THE FOUR LINES OF DEFENCE

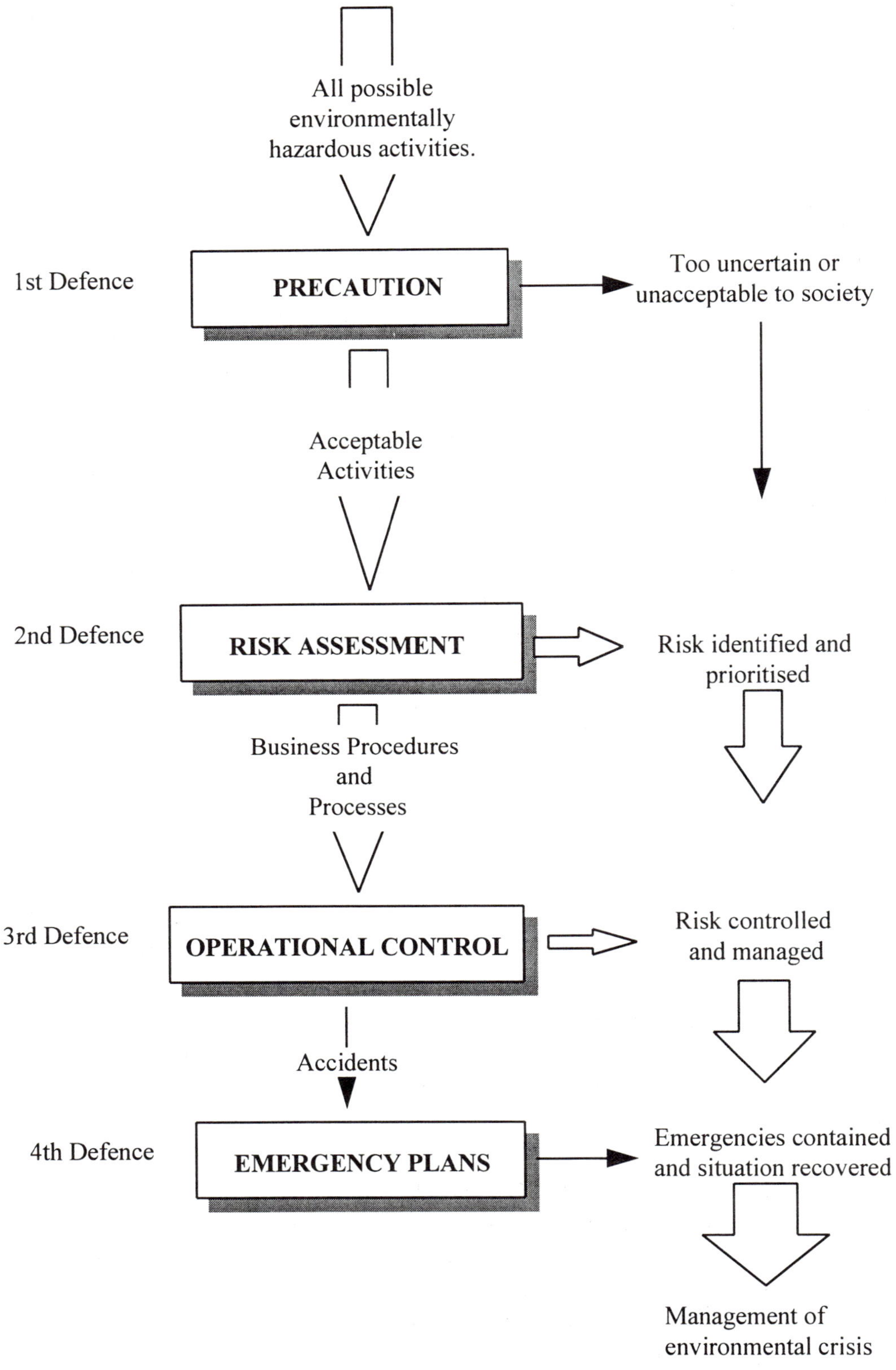

FIGURE 2

PART 2 THE LEGAL FRAMEWORK

3. *Legislation and Regulation*

Just as with our exploration of the environment and environmental management systems, the purpose of this section is not to provide the reader with a comprehensive account of what is a major subject in its own right, but to give an appreciation of the risks that can arise from legislation. This is not to say that legislation is itself a risk, although some may view it as such, but rather that legislation is intended to deal with those matters that society have identified as a risk and that, as a consequence, it wants to control. The distinction may be a fine one but it does illustrate that environmental management can be approached at a number of different levels.

At the lowest level is a system that is designed solely to keep either the company or individuals out of court or to avoid charges levied by regulatory bodies. This is generally termed a compliance based system. It controls risks but only from that narrow standpoint and is therefore to be regarded as the minimum that must be done.

Upon the compliance foundation can be built further elements of the system that go to make a more proactive and strategic approach. The additional elements can cover:

- Anticipating the effects of resource depletion
- Reputation risk management
- Reputation enhancement through demonstrating "green" credentials
- Positioning the company to anticipate future requirements
- Cost saving through review and adjustment of manufacturing or operational processes
- Identification of possible market openings
- Improving quality
- Improving staff morale and motivation

This is not a comprehensive list and the reader could probably find others that may arise from the special circumstances relating to the individual industry. Some of these additional elements are dealt with in later chapters but this one deals only with the driving force behind the foundation level of compliance.

Europe

While Britain has had environmental legislation for many years, there is no doubt that current legislation is largely being driven from Brussels. A number of other northern European countries are particularly active in this field. European Union strategy is defined in a series of environmental action programmes. The current one, the fifth, is the most comprehensive of all and focuses on:

- Anticipating the effects of resource depletion
- Initiating various changes intended to encourage sustainable development
- Encouraging social responsibility
- Using a wide range of instruments such as financial ones, to tackle individual issues.

From this programme a package of further directives, some of which are already drafted, can be expected to emerge to deal specifically with:

- Sustainable use of resources and the setting up of conservation zones
- Integrated pollution control
- Energy conservation
- Urban environmental improvements
- Environmental health and safety
- Improved mobility management

All of which have the potential for significant impact on business.

In addition, various other initiatives are afoot such as the recent setting up of an environmental agency, largely for the purpose of co-ordinating both information gathering and action. While the details need not concern us, the message is clear. The environment is an area that will become increasingly subject to control and direction. Subsidiarity or no, industry and commerce in this country are faced with the prospect of more legislative control and therefore increasing risk from this source, and as a consequence, an enhanced potential for crisis creation.

The United Kingdom

Before looking at the existing legal liabilities arising from national legislation, it is worth considering what remedies exist for legislative breaches or for causing environmental damage. These are the actual costs that a company can face arising from non-compliance and can be considered as a measure of the level of risk that a business faces.

There are three main heads:

1. Criminal penalties arising from a successful prosecution, which may be against an individual, such as a Director, as well as a company, and include both fines or imprisonment.

2. Clean-up costs that are intended to return the environment to its former state. Such costs may be incurred either alone or in conjunction with a criminal conviction and may cover, in some cases, damage that was caused historically. Measures that are required to be put in place to prevent a re-occurrence of the incident also fall in this category.

3. Civil damages, injunctions, stop notices etc., usually brought by neighbours or others affected by the activity who can show that they have suffered harm. This can be interpreted quite widely by the courts and may well include historical activity that was not necessarily caused by the current site or process owner. Although an enforcement notice may not be a direct cost; disruption to, or loss of production, may incur considerable cost irrespective of the implementation of the necessary remedial measures.

Civil liabilities will arise through either nuisance or negligence. The former may be private nuisance, that is unlawful interference with an individuals' rights or enjoyment, or it may be a public nuisance, that is unlawfully affecting the convenience or comfort of a class of persons.

Significantly a nuisance liability can be acquired with the purchase of land that is contaminated. It is the current owner and not the original polluter that is liable. Acquisitions should therefore be subject to environmental audit to ensure that a potential crisis is not being acquired at the same time.

Negligence arises if there is a breach of a duty of care to another person, who could reasonably be foreseen as being affected by the action, and who suffers damage as a result.

In addition, regulatory bodies have wide ranging enforcement powers and failure to comply with a duly served notice is a criminal offence. While enforcement of environmental legislation has been fragmented, recent legislation has introduced the concept of integrated pollution control and the proposed combination of some regulatory bodies into a single environment agency, suggests that a much more co-ordinated approach can be expected in future. Environmental law is developing rapidly both in terms of proposed legislation and also terms of case law, as courts interpret legislation in relation to new knowledge and circumstances.

The best guide to government policy is the almost annual updates that are issued to the 1990 document "This Common Inheritance" and, interestingly the Labour Party's Policy document "In Trust For Tomorrow" indicates that there is a common approach, namely of increasing control, by both major parties. Thus, at national level, like the European level, business can expect tighter regulation in future and would be well advised to position itself now to meet, not just today's requirements, but those that can be reasonably anticipated.

As a cautionary tale it is worth considering the fate of Dairy Crest. An employee accidentally left a tap open on a caustic soda storage tank for about ½ an hour at the company's Aspatria Works. Although the caustic soda was contained in a bund it dissolved a protective coating and escaped through a crack in the bund wall. The crack had apparently occurred as a result of adjacent construction work.

Some 6,000 litres of caustic soda then entered the River Ellen via a surface water drain. The pollution destroyed fish and invertebrates for a considerable distance downstream in what was defined as a first rate salmon river.

The company was prosecuted under the Water Resources Act by the NRA and pleaded guilty before Carlisle Crown Court. The fact that the company could show that design, construction and operations were not defective, that a recent independent "due diligence" survey had failed to identify anything defective and that reasonable precautions had been taken to prevent pollution, weighed with the judge, who imposed a modest fine of £2,000.

Unfortunately for the company, improvements to the plant to minimise the chance of re-occurrence cost about £150,000, while clean-up and compensation costs amounted to another £50,000. In total the costs were nearly 100 times as much as the fine. Thus, the liabilities arising from legislation are substantial and so a breach can constitute a crisis for a company regardless of the actual effect that the action has had on the environment.

Integrated Pollution Control

As already mentioned the significant development that has taken place with the Environmental Protection Act (EPA)1990 is the concept of integrated pollution control. This in effect requires regulatory authorities to co-operate to ensure that all wastes arising from an activity or site are considered together. Thus it is not possible, for example, to avoid liabilities arising from solid waste disposal by simply incinerating the waste and emitting it, or rather its combustion products, to the atmosphere. The regulatory authorities are required to ensure that the best available technique is used to deal with wastes, although at present they are also required to have regard to costs. In dealing with wastes the principles set down in legislation are:

- to control,
- to minimise,
- to render harmless.

The integration of the regulatory bodies themselves is now to be achieved with the Environment Act, which requires the creation of an environment agency to be formed from the combination of the National Rivers Authority, Her Majesty's Inspectorate of Pollution and the Waste Regulatory Authorities. It should be noted that a similar, but separate, situation will be created in Scotland. Whilst this will not in itself create additional liabilities, it points to the direction of future developments, both here and in Europe, of closer regulation. Liability is not confined to current activities. Section 61 of the EPA requires Waste Regulation Authorities to inspect their areas and if noxious gases or liquids are found to take such steps as are necessary to avoid pollution or harm. Costs may be recovered from the current owner of the land. The Government has now issued its long awaited paper on contaminated land. This suggests that the original proposal for contaminated land registers is now abandoned and that a compromise has been adopted. Land should be restored to a condition such that is "fit for purpose" and that the "polluter pays" principle should be applied. However, it is recognised that the original polluter may be long gone and that the responsibility will generally fall on the present land owner. Thus the liability of a Waste Regulatory Authority's reasonable expenses incurred in carrying out a clean up operation of noxious gases or liquids emanating from land that has been used for waste disposal, remains primarily with the owner of the land, not the disposer of the waste.

The summary in the next section gives the main legislation under which environmental liabilities can arise. However, it is by no means definitive. Various other Acts cover more minor issues and the situation in Scotland is significantly different as the EPA does not apply.

Furthermore, as much of the liability can arise from nuisance or negligence under common law, the situation can potentially change with differing interpretations from different Courts. The advisability for professional legal assistance can not be over emphasised when it comes to the precise interpretation of liabilities under the legislation and of the strategy to be followed in defending a summons.

Premier Waste, a Wiltshire based solvent recovery company, has recently been convicted of two solvent discharges into controlled waters. While it intends to appeal against the conviction and it pleaded not guilty in the magistrate's court, it offered no defence. Both the court and the NRA are likely to view this attitude as unnecessarily time consuming and costly and are, as a result, likely to take a dim view of the company's actions. It can therefore expect greater financial penalties than if it took a more reasonable or co-operative approach. Whether a company can or cannot comply with legislation there is little point in increasing the risk or magnitude of a crisis by adopting a perverse attitude.

Key Acts

The key legislation that affects the day to day operations of most companies are fortunately relatively few in number. Listed below are those pieces of legislation that constitute a survival kit, and which must be borne in mind by management:

1. **Water Resources Act 1991**

Sections 85 and 166 deal with the pollution of controlled waters, the definition of which it should be noted includes virtually all watercourses, ponds and lakes. It is an offence to cause or knowingly permit poisonous, noxious or polluting matter or solid waste to enter such waters without a consent. Note that acts of both commission and omission are offences. Section 161 gives the NRA (or the Environmental Agency in the future) power to clean up and recover costs and makes it clear that there is strict liability.

2. **The Environmental Protection Act 1990**

Section 33, which replaces Section 3 Control of Pollution Act, deals in similar vein with solid waste. It is an offence to cause or knowingly permit the deposit of controlled waste or of dealing with it in a prescribed manner, including treating, keeping or disposing of it, without a licence. Controlled waste is defined in the Act but an EC Directive on Waste has altered this by introducing a new definition. However, the effects are very similar in that they both cover most waste products from industrial processes that are not chemically or biologically inert.

Section 34 imposes a duty of care on the organisation in respect of the transfer or keeping of controlled waste. Waste must only be passed to a licensed carrier for disposal at an authorised site. There is a system of transfer notes that must be completed so that each waste consignment has an audit trail from producer to final disposal.

Section 79 relates to Statutory Nuisance, which can be widely interpreted as almost anything arising from premises and that includes land, that is prejudicial to health or a nuisance. Local authorities are able to serve an enforcement notice on the relevant person or the owner of the land. However, an aggrieved individual, such as a neighbouring resident or business, may also have recourse to enforcement under this legislation.

Of most significance is Part 1 of the EPA which introduced the concept of integratedpollution control as part of the control of atmospheric emissions. Section 6 requires persons who propose to operate a prescribed process to obtain a consent, to which conditions may be attached. The Act also introduced the concept of control by means of the Best Available Technique not Entailing Excessive Cost. (BATNEEC).

3. **Control of Pollution Act 1974**

Although the provisions of this Act have largely been superseded by the EPA, section 60 enables local authorities to take action to control excessive noise emanating from work or construction sites. This Act is also still relevant in Scotland which has not adopted the EPA.

4. **Planning (Hazardous Substances) Act 1990**

This extends the powers of planning authorities, whose main concern is land use planning, allowing control over the use or storage of specified quantities of substances capable of causing fire, explosion or toxic hazards to neighbours. Aimed at regulating industrial activities, the Act requires consent to be obtained before operating.

Regulatory Bodies

Currently, regulatory responsibilities are dispersed and this reflects the piecemeal development of legislative control. The proposed Environment Agency will do much to rectify this state of affairs. The main agencies are summarised below and addresses are given in Appendix B.

National Rivers Authority ("NRA") - created by the Water Act 1989. This independent Government agency is about to form part of the Environment Agency and is responsible for enforcing the pollution control provisions of the Water Resources Act 1991 in England and Wales. Enforcement is actually discretionary rather than a positive duty. It also has duties in relation to water resource protection and flood defences.

River Purification Boards ("RPAs") - in Scotland the water pollution control provisions of the Control of Pollution Act 1974 ("COPA") are enforced by the River Purification Authorities and islands councils.

Her Majesty's Inspectorate of Pollution - ("HMIP") - created in 1986 but to be absorbed into the Environment Agency. HMIP has responsibility in England and Wales for enforcing the provisions of the integrated pollution control under the Environmental Protection Act 1990 ("EPA") and the provisions of the Radioactive Substances Act 1993.

Her Majesty's Industrial Pollution Inspectorate ("HMIPI") - holds the responsibilities for administering pollution control under various Acts in Scotland as a part of the Scottish Office.

Waste Regulation Authorities - there are a number of waste regulation authorities throughout the country and they are the final element of the proposed Environment Agency. Waste regulation authorities enforce the waste management licensing provisions of COPA and the EPA 1990. They are generally county councils in England, district councils in Wales and islands or district councils in Scotland, while for Greater London, the London Waste Regulation Authority, for Greater Manchester, the Greater Manchester Waste Disposal Authority, for Metropolitan County of Merseyside, the Merseyside Waste Disposal Authority undertake this role.

Local Authority Environmental Health Officer - the environmental health department generally has a duty to comply with the local authority's obligations under the statutory nuisance provisions of the EPA such as the control of noise. They are also responsible for enforcing the local authority air pollution control provisions of Part 1 of the EPA.

Local Planning Authority - the planning department of the local authority is responsible for development plans and development control including Environmental Impact Assessment. Where the subject of development relates to waste or minerals, the county will fulfil this function, otherwise it will generally be the district council or unitary authority. In an urban development corporation area, the corporation will fulfil the development control function.

Health and Safety Executive ("HSE") - set up by the Health and Safety at Work etc. Act 1974, the HSE is responsible for enforcing health and safety laws. It is the executive arm of the Health and Safety Commission, also established by the 1974 Act. The HSC has the power to propose health and safety regulations and approve codes of practice. It may have separate executive elements dealing with individual industries such as Her Majesty's Railway Inspectorate.

In conclusion, it is important to be aware that the various Acts give regulatory bodies a wide range of enforcement measures. These typically include the service of enforcement or stop notices. Such notices may require, or subsequent negotiations may establish, the obligation for appropriate remediation or clean up measures. Should these not be undertaken then the Regulator can carry out the work itself and recover the costs from the landowner.

It is well worth developing a good working relationship with the Regulatory Bodies. Some can provide very helpful advice and training aids. The NRA's Pollution Prevention Pays initiative helps companies to identify potential problems, and how these can be avoided or reduced. A video covering site drainage, deliveries, storage of potential pollutants, cleaning, waste disposal and emergency actions is available on request.

The main elements of the regulatory framework are summarised in Table 1 below.

Summary Of Regulatory Framework

LIABILITY	REGULATORY BODY	SANCTION	RESPONSIBLE PERSON
Integrated Pollution Control and Local Authority Pollution Control: Part 1, EPA	HMIP/ Environmental Agency	Fine Imprisonment Abatement Measures Clean-up costs	Authorisation holder, person convicted of release
Part 11 EPA Disposal of Waste	Waste Regulation Authority	Fine Imprisonment Clean-up costs Civil Liability	Disposer of waste licence holder
Clean-up of controlled wastes: s. 61 EPA	Waste Regulation Authority/Env. Agency	Clean-up costs	Owner at time of offence
Statutory Nuisance Part 111, EPA	District Council/Aggrieved Persons	Fine Abatement Measures Clean-up costs	Person responsible for nuisance of owner or occupier
Duty of Care: s. 34 EPA	Waste Regulation Authority/Env. Agency	Fine	Any person who produces, imports, carries, keeps, treats or disposes of controlled waste
Pollution of Controlled Waters s. 85 Water Resources Act 1991	National Rivers Authority/Env. Agency	Fine Imprisonment	Person who caused or knowingly permitted pollution
Statutory Nuisance Part 111, EPA	Local Authorities Private Individuals	Fine Abatement Measures Clean-up costs	Person responsible for nuisance or owner or occupier
Common law: Negligence, nuisance, Rylands v Fletcher, Ballard v Tomlinson	Plaintiff	Damages Injunctions	Various

TABLE 1

4. *The Precautionary Principle*

In the foregoing chapters we have briefly considered the complexity and vulnerability of the environment, the likely causes of an environmental crisis, and the regulatory system and authorities that constitute the legal liability that a business must face. At the end of Chapter 2 it was suggested that there were four possible lines of defence against environmental crises. The Precautionary Principle is the first of these.

However, before looking at this in detail it is reasonable to ask why the principle is relevant today and why, more importantly, it will become more relevant tomorrow. The reason has everything to do with the pace and extent of technological change and it has several elements:-

- Numbers of new technologies and products
- Speed of development
- Ability to measure effects.
- Greater knowledge about the environment

New Technologies and Products

Scientific research and industrial technology are constantly introducing an ever larger number of new products year by year, this includes not only chemical compounds, but also their wastes and by-products, which arise from new processes of increasing complexity. This is perhaps a sweeping generalisation but there is no doubting that the underlying trend is away from a relatively small number of "refined products" i.e. ones based on those occurring in nature, towards a relatively large number of entirely man-made products.

While individually these products are tested to standards that seem appropriate by today's knowledge, what is not tested, and cannot reasonably be so tested, is each product against all other products in the various different ecosystems where it may be manufactured, used and disposed of. Thus the interaction between these products is poorly if at all understood. There is evidence that within many ecosystems the total load of pollutants is building up and adversely affecting the ability of the system to adjust to new disturbances. We do not need to postulate some catastrophic reaction between two or more products causing a crisis. The steady build up, perhaps by way of legally permitted discharges, of the total load of pollution may cause a sudden and irreversible breakdown in an ecosystem.

Speed of Change

One very good reason why we could not possibly test all products in all circumstances is that it would take far too long to do so. Commercial and other pressures dictate that products are introduced in the shortest possible time, not the longest. Here we are trapped in a vicious circle for not only do we need to introduce new developments quickly but we are attempting to introduce more each year. The net result is that in absolute terms our testing, in spite of increasing efforts and sophistication, is not improving but in effect becoming more restricted year by year.

Two examples illustrate the difficulty here. The first is the concern over female hormones, or chemicals that mimic female hormones in the environment, which appear to be reducing male fertility, affecting both ecosystems and man. Such an effect was not suspected when manufacture of the hormones themselves was contemplated; let alone the property of a diverse range of chemicals to mimic hormone action.

Furthermore, it has taken many years for us to become aware of the possible link, and to prove it conclusively may yet take several years further work, by which time very serious damage may have been done.

The second example is the catalytic converter now fitted to all new petrol engine cars to reduce nitrous oxide and other emissions. Introduced rapidly, for the best of intentions, it now emerges that the pattern of car usage in urban areas reduces its effectiveness considerably. Most journeys do not allow the converter to reach its optimal operating potential. Additionally the life cycle costs in environmental terms were initially underestimated so that the ecological cost/benefit ratio is not as favourable as it was predicted to be. Here we have a quick response to a relatively "new" problem, the air pollution caused by the motor car, that is now seen to be less than optimal with hindsight.

Measurement

Conversely to the pressure to reduce timescales, and somewhat perversely,year by year we are able to detect and measure ever smaller quantities of an ever wider range of chemicals. This is due to greater sensitivity of detection devices. Like crime, the number of pollution incidents is increasing in part due to better reporting; in other words the more we look for, the more we find. We are now much more aware of exactly where our products are going in the ecosystem, even if we are not certain of what they do when they get there.

Lest all this seems of purely academic interest, an example, only one of many, shows what may be happening on a wide scale. Recent research in America has concluded that the commonly used pesticide often found in drinking water, atrazine, is a probable carcinogen. As its effects have been unknown, EU legislation has set a precautionary limit of 0.1 mg/l (one ten millionth of a gram per litre) in drinking water. Both the water industry, which has faced costly investment to meet this limit and the pesticide manufacturers have long argued this limit is too low. It now appears it may be too high and the EU precautionary approach has, perhaps, been amply justified.

Look Before You Leap

This then in essence is the precautionary principle, one that is equally applicable to society as a whole, business and individuals and to the whole range of business activities, not just their relationship with the environment. When dealing with a complex and dynamic situation it is better to err on the side of caution in respect of actions whose consequences are unknown, or only partly known. While this is intended primarily to avoid long term damage to the physical environment it is very difficult to divorce it from other issues such as health and safety.The current environmental management systems imply, if not directly state, that companies should exercise precaution in their operations and in the projects that they propose.

The principle is generally accepted by the Government and the EU. Article 130v of the EC Treaty of Rome states that community policy on the environment must be:

"based on the precautionary principle and on the principles that preventative action should be taken, that environmental damage should as a priority be rectified at source."

National governments are therefore obliged to follow this principle in formulating policy, although the Council of Ministers made up of national government ministers is the body responsible for deciding how this should be implemented.

In this country, government strategy was defined in the White Paper, This Common Inheritance, Britain's Environmental Strategy and in the UK Sustainable Development Strategy. The latter argued that:

"when potential damage to the environment is both uncertain and significant, it is necessary to act on the basis of the precautionary principle."

This was described in the 1990 White Paper in the following terms:

"Where there are significant risks of damage to the environment the Government will be prepared to take precautionary action to limit the use of potentially dangerous materials or the spread of potentially dangerous pollutants, even where scientific knowledge is not conclusive, if the likely balance of costs and benefits justifies it".

The policy is that the threshold for a proposed action (development) should be raised in the absence of conclusive scientific proof of lack of risk. Thus the absence of proof of likely damage is not sufficient to justify the proposed action but it does face the proposer with the difficult task of proving a negative. However, it is not all risks that concern us but only those that are "significant".

This rightly throws the responsibility for invoking the application of the precautionary principle on those that can show that there is a significant risk, as opposed to showing that there is any risk at all. Clearly such a principle could potentially be misapplied so that development and innovation was stifled to the detriment of business, the economy and society as a whole. Like most things to do with the environment, a balance must be struck; however difficult it is to decide where that balance should be.

An example of this is the current debate over the effects of electro-magnetic radiation emanating from the national and domestic electricity supply systems. Some scientific studies have suggested that there is a link between childhood cancer and exposure to high levels of this radiation. However, this evidence is open to considerable doubt so that "significant" risk cannot be said to have been proved. Thus the onus is not yet on the electricity supply industry to prove conclusively that the threshold for taking action should not be raised. Nevertheless it is no secret that the industry, for a variety of reasons, seeks to site supplies away from residential properties, where this is physically possible and commercially justifiable.

The precautionary principle is particularly relevant to projects. British Gas Exploration and Production for example passes a proposed exploration programme through the stages of:

- Site appraisal
- Baseline survey
- Environmental Impact Assessment
- Contingency planning and Emergency Response.

Precaution can be exercised at the site appraisal stage and the programme aborted or modified if uncertainty is too great. However, if the proposal is allowed to proceed, environmental impacts (risks) and emergency plans are integrated into the process. We shall go on to consider these aspects in succeeding chapters.

As a final twist, the precautionary principle must also apply to corrective action that is taken as part of any crisis management. The desire to "do something" in some circumstances may easily lead to actions that compound the problem and it may be better to do nothing in these circumstances. The Rio Declaration on Environment and Development, 1992, says:

"where there is significant or serious or irreversible damage, lack of full scientific certainty shall not be used as a reason for postponing cost-effective measures to prevent environmental degradation".

In other words take all necessary measures to stop the situation deteriorating but be very cautious about trying to restore the environment.

Effect On Business

The precautionary principle, when applied to business, may be said to be nothing more than prudent management. The adoption of an environmental management system for example as outlined in Chapter 2 could be considered as precautionary. However, the principle is mainly applicable to, at one end of the scale, the introduction of a new process or product, and at the other, to the development of a new market or manufacturing installation. This is particularly so when the potential environmental uncertainties and liabilities are great, such as when critical environmental processes are affected.

The actions that should be taken in such cases can include a combination of:-

- Greater investigation
- Greater safeguards
- Increased monitoring.

Greater investigation might include increasing the number of tests carried out, widening their range or prolonging the testing period. It might also include additional investigation of the ecosystems or of the environmental functions that are potentially at risk. Time might profitably be spent carrying out a literature search to see if comparable or related studies have been undertaken by others for example. However, it is recognised that greater investigation comes at the most sensitive stage, in financial terms, of a product's life-cycle. Substantial costs may already have been incurred and the danger of being beaten in the race to be the first in the marketplace are unlikely to make this a favoured precautionary option late in the development process. However, it must be the favoured approach at the project feasibility study stage.

Greater safeguards built into the production, distribution or any disposal service are also going to cost money but need not substantially delay the introduction of the product or the plant. Indeed, if there are substantial risks then such precautions may only anticipate increased regulatory requirements and may be a sound capital investment at the outset. If the greater safeguards relate to plant management then the costs are going to be spread over the life cycle of the project. The safeguards may in any event be financial rather than operational or physical.

Increased monitoring, intended to give early warning of impending trouble, is in many ways the cheapest option. Unfortunately it is also the least effective, at least if implemented alone. By definition it will only detect when something is going wrong. While monitoring may mean that you may have more time to react to an incipient crisis, it would be better to take preventative action and avoid the crisis altogether.

In coming to a decision about precaution, a company is seldom likely to have total freedom of action. Options may be curtailed by financial institutions exercising due diligence before advancing funding, regulatory authorities controlling various aspects, shareholder, neighbour, or public attitudes. One option is to consider pilot schemes, which still offer the chance of being first in the marketplace, combined with market testing but with limited financial and environmental exposure.

To assist in such decision making, it is unlikely that techniques such as conventional cost/benefit analysis will be much use. "Costing" the environment, while by no means unknown, is a very imprecise science. What is more it follows that, if the precautionary principle is applicable, then by definition there is a great deal of uncertainty over the factors that must be evaluated in the analysis, which makes quantification of the costs very difficult.

Uncertainty, by its very nature, cannot be evaluated. Of more use, indeed more used, is environmental assessment, which is mandatory for certain types of land or physical development and those likely to have significant environmental effects. Environmental Impact Assessment (EIA) tends to apply to land-use planning rather than product or process planning and so forms part of the town and country planning system. The procedure brings the analysis of likely impacts and the measures that can be used to mitigate adverse impacts into the public domain. The decision on whether to permit the development may be taken by the local planning authority or, more likely, by the Secretary of State for the Environment, (who is advised by the Inspector appointed for the purpose) after a public inquiry. Either the local authority or the Secretary of State may choose to apply the precautionary principle, if there is uncertainty about significant issues, and refuse permission for the development. This effectively takes the decision to apply the principle out of the hands of the business that is seeking permission to develop.

However, business may have to accept the fact that to precede with a new and little understood product, with possible effects on fragile or particularly sensitive ecosystems, will represent an unacceptable liability. In such cases the only sensible precaution is not to proceed.

Part 3 RISK AND ITS CONTROL

5. *Risk Management*

Throughout preceding chapters the word "risk" has kept arising and the need to identify and control risk is a key component in the management of crises. It is a key because at times it enables the potential causes of crises to be avoided; the most effective management technique there is. At worst it is still helpful in providing an insight into the likely nature of the impending crisis. Remember, however, that we are here considering risk in relation to the business arising from its interaction with the environment. This is not the same as the risk to the environment arising from the business, although the two are closely related.

Risk analysis and closely related techniques such as risk assessment, risk management, hazard and operability studies (HAZOPS), failure mode and effects analysis (FEMA), all form a huge field of study to which this book cannot hope to do justice. We will, therefore, look at the principles that are being applied in such studies and develop an extremely simple system that companies may use themselves, more as a guide as to when to seek professional help than as a definitive technique to be applied in any particular instance.

Given the very wide definition that is now accepted for the environment, few if any business activities can be envisaged that do not have some sort of environmental effect associated with them. Given also the potential liabilities, possible hazards and implications then there is almost certainly a risk to the business. We have already suggested that risk is different to uncertainty, in that it can, to a greater or lesser extent, be quantified. It is the probability of harm being experienced to the balance sheet, either through direct costs, physical assets or the reputation of the business. The inter-relationship of risks and costs can be summarised in the drawing in Figure 3.

THE RISK/COST RELATIONSHIP

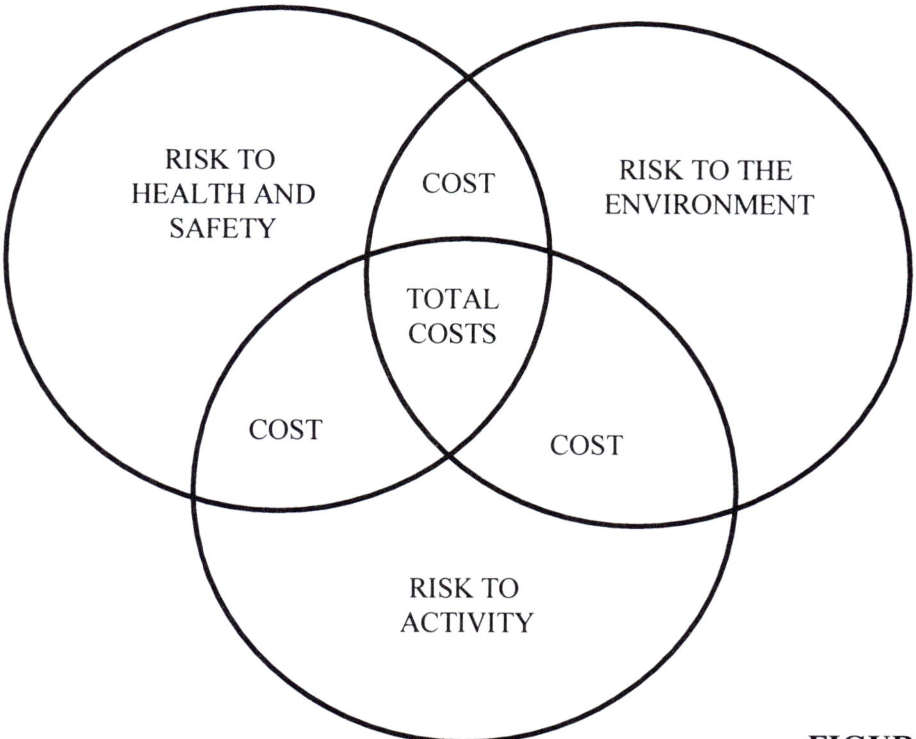

FIGURE 3

The planned and systematic management of risk is necessary, not just to deal with crises, but for sound business planning. It can give some control over strategy, resource allocation, operational matters and also the interaction of the business with its stake holders. It can be very helpful in reassuring management that its corporate and personal liabilities are under control. Curiously, however, the current management systems are silent on the issue of risk assessment but this should not inhibit a business from including it in their own system.

As we shall explore in more detail below, there are three main elements to risk management:-

- Identify the risks
- Analyse their magnitude and probability
- Identify their likely consequences on the business.

Knowing these matters then it is possible to set about reducing risk and hence reducing potential crises. Reduction can be achieved by using physical measures, administrative procedures or financial planning. The separation of these elements may be somewhat academic as there is considerable overlap between them. For example re-engineering a process may have a capital cost, revenue implications, such as energy savings and require new procedures and training, all of which have financial consequences.

However, it is possible to finance risk without the other measures. Broadly this means taking steps either to accept and live with the risk, transfer it to another party by means of a contract or insurance or to make provision for the financial consequences of risk by means of bonds or financial reinsurance. In this case the potential crisis may be identified but no steps taken to avert it. While this is a potential strategy for crisis management it is not the subject of this book, which is concerned with averting or controlling risk itself.

Whatever risk management plan is adopted, there is a need for regular review to establish that it remains relevant and that it is delivering the required performance. There are numerous changes, which can be either internal to the company or external, such as products, markets, legislation and technology, that might invalidate the plan.

Auditing and performance review will help both to assure management that its liabilities are under control and in making future decisions. This is, of course, a necessary part of any management system including environmental management such as BS 7750 and crisis management.

Before proceeding further it is necessary to define clearly the terms commonly found in risk assessment. The crisis will probably start with a <u>failure</u> which is the deviation from the design or operation of the plant, process or product. This may lead to an <u>environmental hazard</u>, which is a situation that could lead to the degradation, either directly or indirectly, of the environment whether in the short or long term. The hazard has therefore the potential to lead to a <u>crisis</u>, which is an uncontrolled or unplanned event giving rise to environmental consequences with significant cost implications. The <u>business consequences</u> are the outcome and may range over legal action, resulting in fines or imprisonment, clean-up costs, loss of reputation etc. The <u>environmental consequences</u> may be even more diverse and have been outlined in Chapter 1. The <u>severity</u> of the crisis is a measure of its size or magnitude. <u>Environmental Risk</u> encompasses these and is a measure of the potential threats to the environment. It is the combination of the probability or frequency of a damaging event occurring and the severity of the consequences.

Risk assessment was devised to deal with issues of health and safety, and has relatively recently come to be used for the environment. As a result it tends to be concerned with the effects on humans rather than the environment itself. However, the basic techniques are readily transferable, with one exception. This concerns quantification of severity and arises from two main sources. The first we have already dealt with, namely the complexity of natural systems which makes it difficult to be confident of the magnitude or extent, both geographical and temporal, of any accident. The second relates to " valuing the environment". In conventional risk assessment it is not difficult to put a value on physical damage or even a nominal value on an injury or a human life. But what is the value of a habitat, or a species or a view? Considerable ingenuity has been demonstrated in trying to ascribe values to such matters, but ultimately they may be beyond valuation, if not beyond value. This need not, however, invalidate the application of risk management to the environment.

Risk Assessment Method

In simple terms the technique aims to find out what can go wrong, what then happens, how often this is likely to occur and does it matter to the business. If it does matter then there is the potential for a crisis and the appropriate management methods need to be put in place.

The method outlined below is intended to allow management to understand the process, assist them to identify when it can be of value and to communicate meaningfully with professional assessors. It will also enable simple, self-help assessments to be undertaken in order to facilitate management decisions. The effort that needs to be put into the work should relate to the reason for undertaking the assessment in the first place and to the magnitude of the potential risks or significance of the results. For this a simple, do-it-yourself exercise using conservative assumptions and worst case data may suffice. If such an exercise shows, for example, that the risks are not significant, then it is sufficient for its purpose. If it were to identify significant risks then further work, probably by a professional assessor, would be advisable.

Hazard Analysis

The analysis of hazards requires that firstly all possible (but realistic) undesirable events are identified and for this a relatively simple checklist may be all that is required. Secondly, the means by which they might occur to be identified. This can be provided by the more sophisticated techniques of Failure Mode and Effects Analysis (FEMA) or Hazard and Operability Studies (HAZOPS), which are particularly designed for dealing with plant and equipment. Hazards arising from a product are somewhat different. Hopefully these are addressed in the design and testing stage, through mechanisms such as Type Approval or Licensing.

For a rough and ready exercise a comprehensive checklist may suffice for both identification and cause but for greater accuracy and confidence, together with the added value or identifying potential operational but non-critical problems, then a more structural approach is much to be preferred. Usually this requires an experienced chairman and a small multi-disciplinary team familiar either with the whole issue or with individual elements of the site, process or plant. The chairman's role is critical and such a person may have to be found from outside the firm. Sufficient time and resources need to be made available for the team to function effectively.

The purpose of an exercise such as a HAZOP is to explore undesirable deviations from the normal operational condition of the plant or process. Each element of the system e.g. transportation of raw materials to the site, their delivery, storage etc. through to disposal of waste and despatch of the product, are considered in turn. A series of guide words cover all possible failures e.g. too much, too little, reverse flow etc. and, for each deviation, the possible causes and consequences are documented and either possible safeguards or the need for further work are noted.

For a major new plant it will readily be understood that such an exercise can involve many man-hours of study of complex pipework diagrams and component specifications. It can, however, also show up design shortcomings and operability problems that if addressed at the design stage, can save considerable time and money later on. Environmental hazards can be explored at the same time or flagged up for a separate study, for example:-

GUIDE WORD	CONSEQUENCE	RESULT	ACTION
Too much heat	Expansion of reactants/pressure relief valve lifts	Uncontrolled emission	Environmental Manager to advise

Magnitude Assessment or Severity

Having identified how and why a hazard may arise the next issue is what will the effects be? Assuming that the most likely problem will be an uncontrolled emission but that the release might be something other than a "substance", i.e. noise, visual intrusion, radiation etc., then the emission must be tracked through:

- Transport
- Dispersion
- Uptake
- Transformation
- Breakdown

as appropriate, and the consequences identified. Usually it is sufficient to identify the "target" or the final depository of the substance in the environment. This could be plants, water, soil or air but often and more significantly, it will be concentrated as it passes through the food web to end up in the keystone species, quite often man. In many cases it may be sufficient to simply identify the legislation that will be breached.

Complex, quantified assessments are ideal but may be neither possible because of lack of time or information, nor are they always necessary. Simple, qualitative assessments based on expert opinion may be perfectly adequate for enabling appropriate management decisions to be taken. Thus the consequences may be identified as falling somewhere on a scale such as:-

Simple Severity Guide

VALUE	MAGNITUDE	COST GUIDE
1	None predicted	None
2	Minor/limited extent	Normal operating cost
3	Moderate/local problems	Significant additional cost to local plant/side product
4	Severe/widespread	Significant additional cost to operating division/company
5	Catastrophic/transboundary effects.	Cost such that Division/Company unlikely to recover.

TABLE 2

When undertaking a risk assessment, it is usual to err on the side of pessimism and take a 'worst case' to allow for unknowns. This is particularly so in dealing with environmental consequences, which are rarely well understood. The cost guide may help in coming to a judgement on the value. Remember to include fines, repairs, clean-up and reputation in the assessment.

While this may appear simplistic, the benefits arising from a much more detailed study, where there is a limited understanding of systems and limited data on which to work, may not justify the cost. Even this level of analysis may be unnecessary if management is concerned only with strict compliance with regulations. In this case a release either complies with any Integrated Pollution Control (IPC) consent or it does not, in which event it is therefore liable to court action and clean up costs.

Probability Analysis

The likelihood of the event occurring in a given time period is the outcome of frequency analysis. This relies heavily on historical data which, for most risk analysis of industrial plant and processes, is adequate and available. Unfortunately historical data for the environment may not be available, or if it is, it may be derived from another situation with different effluents, ecosystems etc. so that it is not directly relevant. Usually frequency analysis makes use of logic diagrams and fault trees, which can identify how often an event may occur that comprises a series of preceding events, for which the individual frequencies are known.

Nevertheless, as for consequence assessment, a detailed assessment of frequency may not be strictly necessary and a simplified classification such as the one below may suffice. Using a time period of a year and assuming that a frequency of up to once a year is a reasonable benchmark then an analysis might determine values of:-

Simple Probability Guide

VALUE	FREQUENCY
1	Remote (< 10^{-4} per year)
2	Very unlikely (10^{-4} to 10^{-2} per year)
3	Unlikely (10^{-2} to 10^{-1} per year)
4	Likely (10^{-1} to 1 per year)
5	Very likely (1 to 10 times a year)
6	Frequent (10 or more times a year)

TABLE 3

When a range of different events are identified, each with a likely frequency under a given set of environmental conditions, the outcome is known as Probabilistic Risk Assessment. PRA can be a sophisticated tool but one that can be reduced to simple essentials that can readily be applied to management decisions. A figure may be derived, which ranks the relative risks, by multiplying the consequence severity (Table 2) with the likelihood (Table 3). For example a severe/widespread consequence (Value 4) which is very unlikely to occur (Value 2) scores 8. This might usefully be compared with a quite minor event (Value 2) that is likely to occur two or three times a year (Value 5) thus scoring 10. Attention on the latter issue is more easily justified as a priority in comparison with dealing with the former. However, it should be noted that this result may not accord with public risk perception.

This basic analysis will give a reasonably comprehensive picture of what could go wrong, how often it will happen and what the consequences would be. However, in order to arrive at this position, it is likely that a number of assumptions will have had to be made at various stages of the process. Where there is a high degree of uncertainty it would be advisable to carry out a sensitivity analysis to determine how robust the predictions are to changes in the underlying assumptions or even, for that matter, to the data that has been used.

If the results of the assessment are overly sensitive to the assumptions made it would be advisable to attempt a more detailed study before basing management decision on the outcome.

Consequence for Business

Having arrived at a clearer picture of the likely risks, management has available a number of other techniques to assist in determining the best course of action. Risk Benefit Analysis (RBA) considers the benefits that can accrue from managing risk. It starts from the basis that if no action is taken what are the likely costs of the risks that are accepted, and compares this with the benefit i.e. saving the cost preventative measures. If the costs are less than the benefits, then no action is justified, or conversely if the costs are greater then action should be taken. This is inverse of cost benefit analysis (CBA), with which most readers will be familiar. The main difficulties lie in dealing with very unlikely but extremely damaging events and also in assuming a value to certain environmental factors which may be at risk.

The general relationship of the two parameters are shown in figure 4.

Relationship Between Risk and Cost

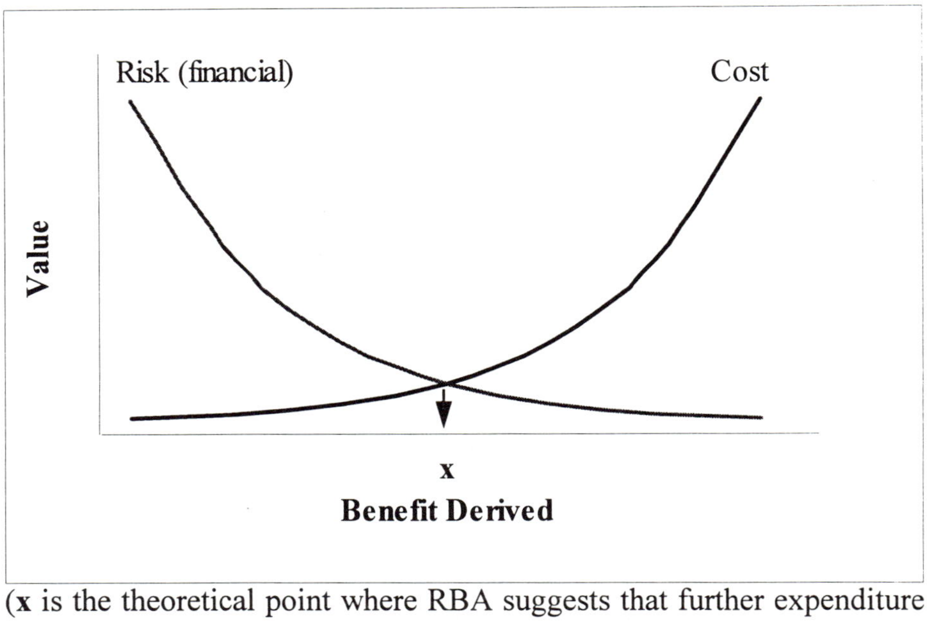

(**x** is the theoretical point where RBA suggests that further expenditure on controlling risk is not justified.)

FIGURE 4

Acceptable Risk Analysis (ARA) tries to overcome these criticisms by avoiding precise valuations and asking the question, "How safe is safe enough?." in answering this CBA or RBA may play a role but are supplemented by expert opinion, company policy, past experience, public opinion etc. in an attempt to reach a consensus view. This appears admirable in theory but in practice it is fraught with difficulty. Not only will there be a whole range of views on the question but the management is now faced with deciding the weight to be accorded to each. There may be a tendency to err on the cautious side which is tantamount to passing the issue back to the previous stage - the precautionary principle. In addition the technique is better suited to controlling the risks of human deaths or injury rather than that protecting the environment. Nevertheless, this is the underlying principle of ALARP (As Low As Reasonably Practicable) which is the principle that the Health & Safety Executive usually adopt. This is illustrated in Figure 5.

THE ALARP PRINCIPLE

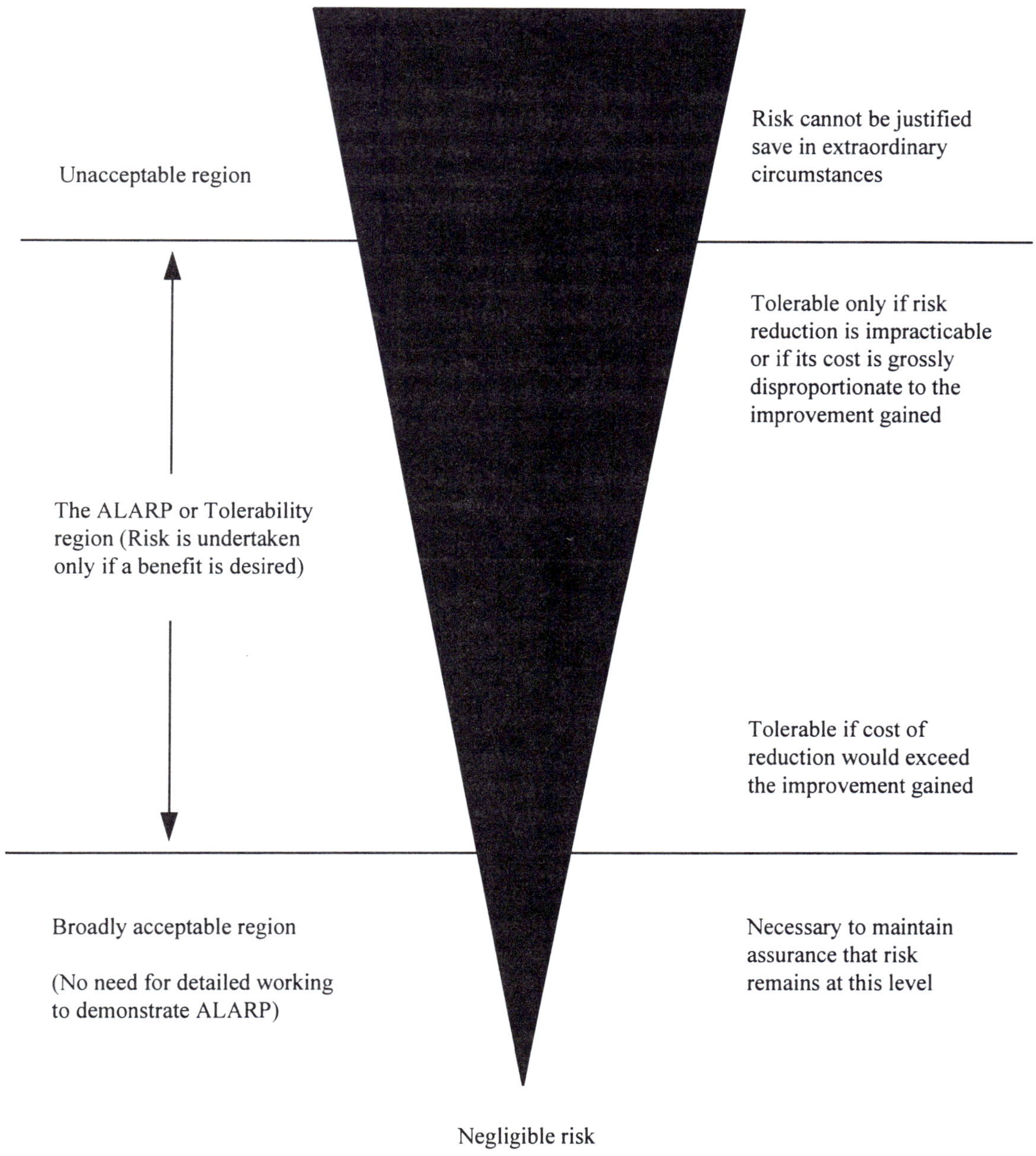

Source HSE

FIGURE 5

Decision Analysis (DA) recognises the difficulties inherent in making value judgements in decisions. It is likely to be of more use to those with regulatory or consent granting roles, rather than business management, as it looks at the possible outcomes, their frequency and the utility (to society) to be expected from each one, rather than the disadvantages. Thus it concerns itself with the use of the plant or product as compared with the risks associated with it. DA focuses attention on who makes the decisions and why; rather than what can be done to minimise company risks, although the two may be closely related. For example, where there is an inherently risky project a secondary project may be advisable to reduce risk or strengthen environmental safeguards.

Risk Perception

One difficulty faced in both risk analysis and in reputation management is the general public's perception of risk. Unfortunately this is seldom in agreement with the mathematically calculated risk associated with any particular activity, and it also varies from individual to individual. Like a crisis, risk assumes a much greater importance if it is perceived to be "close to home".

In general a number of factors affect perception. The following ones tend to reduce the perceived level of risk:

- Familiarity with risk
- Distance
- Having some degree of control over the risk
- The risk occurring at some distant future date
- The risk relates to a relatively large number of small events

While the following tend to cause the perception of the risk to be magnified:

- Unfamiliar, unknown or unseen
- Physical proximity
- Lack of any control
- Immediacy
- The risk relates to a few or a single catastrophic event.

Thus for an individual in this country the chances of dying at the wheel of their own car and close to home are relatively high, but are readily accepted. However, the chance of succumbing to a new and deadly virus from Africa is substantially lower but are much more likely to cause concern and even hysterical press reporting.

There are two significant points to emerge from this for managers. First, before a crisis, the public may be unwilling to be concerned about, or even co-operate over, risks that they believe to be low. Conversely they are likely to be over-concerned and obstructive over big perception risks, even if it is not justified. Secondly, during the crisis, it is pointless and perhaps even counter-productive to try and argue about the real, mathematical value of a risk. The public perception will be of a huge and continuing risk and it will not be amenable to logic, as Shell found out with the Brent Spar.

The difference in perception can, to an extent, be explained by the different viewpoints. The public is likely to be concerned about the risk the company poses to the environment. The company, however, is more likely to be concerned with the risk the environment poses to itself.

Managing Risk Assessment

From the above it is clear that while a simple risk assessment could be undertaken on a self-help basis, which we will look at in Chapter 10, the employment of specialist consultants will be necessary for large scale or complex projects. Assuming this is the course followed, what should the manager be asking for in the brief and be looking for in the final report and how can he assess the significance of the findings and relevance of the recommendations? Even if the exercise has been conducted on a self-help basis, the following checks should help to give some reassurance as to the confidence that can be placed in the outcome.

1. The structure of any report must reflect the key elements of:-

   - What are the hazards
   - What are their magnitude
   - How frequently will they occur
   - What are the consequences and
   - What should be done about them.

2. Much of the technical manipulation of figures will be too time consuming to check and, in any event, should not contain simple arithmetical errors that one could hope to spot. However, the inputs to the process should be checked i.e. where did the data come from, what confidence can be placed in its accuracy together with any underlying assumptions in the process, for example that the plant will never be run over the weekend and tests will always pick up faults? In addition the outputs need to be checked to ensure that consistent units are being used and that like is being compared with like, for example is a once in one hundred year incident a release, a prosecution, a death or something else?

3. Returning to the concept of the black box, the manager needs to satisfy himself that:

   - The quantities, nature and properties of the inputs have been accurately identified.
   - That the parameters within which the process operates have been accurately recorded.
   - That there is no possibility of going outside these parameters.
   - That start-up, close down, maintenance and abnormal operating conditions have been identified.
   - That all outputs, products, by-products, and wastes are identified along with their essential properties.
   - That the movement of goods or materials into or from the black box are covered.

4. Hazard identification is clearly a key element and not one the general manager can easily check. However, do the hazards feel right, are health and safety hazards clearly identified separately to environmental and what methods have been used to identify them?

5. Operational factors are something that will be looked at more closely in the following chapter but the human factor is one that must be included in risk assessment. Both extreme standpoints are dubious, namely that the operator will always function as required or conversely, that the operator will always fail. The fact lies somewhere in between and is affected by variables such as motivation, training, workload etc. What assumptions have been made on operational factors and are they too optimistic or pessimistic?

6. Recommendations should include a list of the options that have been considered and the reason for accepting or rejecting them rather than a take it or leave it choice. The recommended option should also be comfortable to management both in terms of the credibility and company policy and practice. If it does not feel right it is worth trying

to find out why it is so. Do the findings of the study accord with common sense and historical experience?

Objections to Risk Assessment

Hazard analysis and risk assessment are not free of criticism. Charges of:

- They lead to expensive unnecessary work
- Too little reliable data available for the study
- Some hazard and environmental values cannot be quantified so invalidating the rest of the study
- Business decisions cannot be based on numbers alone

are a selection of the ones most commonly heard.

All these have an element of truth but, as this book makes plain, risk analysis in only one of the defences against environmental crises. The technique brings an element of rigour and quantification to what would otherwise be uninformed gut feeling. Costs may be substantial but this is no reason for not asking what they are before making the decision as to whether to accept them or not. Data and figures may well be difficult to capture but even simplified analysis of those parts of the system that are problematic may give reliable "order of magnitude" costs, which assist in decision making. The assessment does not make decisions of its own but rather enables management to make better decisions. However, if the exercise does throw up an inescapable commitment to action, then better it had done so than that management proceeded in ignorance until disaster struck.

Asking the Question "So What?"

Having gone through the exercise of assessing risks, what has the management got to do to make use of it in crisis management? By identifying the size and nature of the risk that arises from each part of the plant, process or individual products, better use can be made of resources to reduce potential crises in either magnitude or frequency or both. Techniques may include any of the following:

a) Design changes before construction,
b) Safety modification to existing plant,
c) Operational changes covering materials storage, handling, training etc.,
d) Changes in processes, raw materials, products etc.,
e) Emergency response plans,
f) Providing for risks through financial instruments such as insurance,

and in extreme cases, a review as to whether the company needs to be operating in this area at all.

The next chapter looks at issues a) to e) in detail and considers how hazards such as that faced by Texaco can be minimised. Early in the morning of July 24th 1994, Texaco's Pembroke oil refinery was struck by lightning, which caused a number of small fires and a shut down of the plant. The fires were rapidly dealt with and the production restarted.

However, at 1330 the plant suffered a massive explosion that was audible 30 miles away. It was perhaps fortunate that it was a Sunday so that relatively few workers were on site and there were only 26 injuries. The resulting fire was not brought under control until the next day and damage to the plant was extensive. Fortunately environmental damage was limited by measures such as bunding, which prevented uncontrolled discharge of contaminated fire fighting water. It appears that the cause of the accident was the disruption to the production process caused by the initial shut down and restart, which gave rise to abnormal flows in part of the plant. This leaked flammable fluids that vaporised and then ignited.

Lightning strikes are unlikely events but not an insignificant hazard for a large site with flammable materials. Curiously though, it was not the lightning that was the direct cause of the accident but rather the abnormal restart operation. Since this could have occurred after incidents other than lightning, the chance of it occurring must have been much higher and it is, in any case, specifically mentioned in legislation and guidance.

This illustrates both that the risks that can arise from an event may be very substantial and that, while the course of events leading to an incident may be convoluted, the hazard is usually foreseeable. Furthermore, it illustrates that an accident seldom has a single cause but is usually the result of a series of incidents each one of which might have been preventable and the control of which would have prevented the crisis. It is the purpose of operational control to ensure that the chain of events is not started, or if it is, it is quickly broken before the crisis occurrs.

Insurance

While it would be unwise to provide for risks solely through insurance, always assuming that it is available, passing some of the risk onto another party by means of insurance is a prudent measure and a reasonable element in crisis management. Unfortunately the environmental insurance market is still poorly developed and coverage limited.

Public liability policies are confined to incidents which occur during the period of insurance and specifically exclude gradual or past pollution. The exclusion is intended to prevent coverage of the typical pollution incident, which is long term and gradual, such as where the on-site contamination is historical and slowly released onto neighbouring land or the local environment.

In addition, most policies are "third party" that is they only relate to damage caused to neighbours and others. They do not cover losses to the insured arising from damage to plant or processes or to cost arising on the insured's own land. This is in spite of growing interest in environmental insurance that has seen the introduction of some new policies that cover

- Own site remediation as a specific issue
- Off site remediation costs
- Environmental impairment liability costs.

These new policies are intended to cover the liabilities that arise at site acquisition, rather than companies that have been operating for a period on a site and now find that, for whatever reason, their risks are insupportable alone.

When policies are designed to cover unexpected pollution incidents, the insurer will require an environmental audit to be carried out before accepting the risk. If the insured is carrying out operations on the site which may, potentially, cause contamination, then the insurer may insist on periodical audits covering both the condition of the site and the operator's environmental management system. These preconditions can be extremely expensive and, of course, the premium will be commensurate with the risk, always assuming that the insurer is prepared to cover the more significant ones.

Conclusion

From this discussion it should be clear that the business needs to have a systematic and planned approach to risk management, integrated with general managerial decision making. Risks are particularly associated with change, which needs careful control. A corporate culture that "risk aware" should be developed and this is assisted if there is a quality culture that emphasises individual responsibility, the ownership of problems and no blame for their identification and resolution.

6. *Operational Control*

It is but a small step conceptually from risk assessment to the third line of defence, operational control, although in reality much time and effort may be required to make the transition. The closeness of the relationship has been emphasised by Her Majesty's Inspectorate of Pollution (HMIP) with the recent launch of risk ratings for integrated pollution control processes, known as Operator and Pollution Risk Approval (OPRA).

This has been introduced primarily to enable HMIP (and presumably the Environmental Agency to be), to make best use of its inspection resources. However, it holds out the promise of reduced subsistence charges for operators who can demonstrate improved performance. The method has two components: an evaluation of the operator's performance and an appraisal of the hazards of the operation. For each component seven elements are assessed.

Operator Performance

- Compliance with limits and adequacy of records.
- Knowledge of authorisation requirements.
- Plant maintenance and operation.
- Management and training.
- Procedures and instructions.
- Frequency of incidents and justified complaints.
- Auditable environmental management systems.

Pollution Hazards

- Hazardous substances.
- Technology for minimisation.
- Abatement technology.
- Scale of process.
- Location relative to sensitive targets.
- Frequency of operation.
- Use of offensive substances.

On the pollution side we find the hazards themselves and factors such as frequency, very much as we might expect. However, we also find "minimisation" and "abatement," which are operational aspects. On the performance side comes "maintenance and operation," "management and training," "procedures and instructions," "frequency" again and last but most significantly "auditable environmental management systems" (EMS). Here is a regulator derived system in which hazard assessment is intended to drive operations and performance. The message could not be clearer. Operational control is the means of managing risks and, risk assessment will become a necessity, not an optional choice.

This also points to a potential tension in operational control namely a conflict between good environmental management and commercial goals. In order to meet environmental objectives, whether for the management of crisis or any other matter, then costs may be incurred in the design of plant, operating procedures or a number of other ways, which will run contrary to financial objectives.

It is not the place of this volume to deal with the economics or environmental management and readers are referred to "Improving Environment Performance: A Proven and Effective Approach" in this series for an exposition of the issue. Suffice it to say, most companies that have instituted a well thought out EMS have been able to demonstrate off-setting cost savings, as a result of the exercise, that have resulted in rapid pay back and long term gains. Our consideration of the issue of risk has demonstrated that investment in environmental crisis management through risk minimisation or preferably avoidance, is akin to the philosophy of insurance. It is another method of loss control and offers the opportunity, for a small expenditure now, of a large saving in the future. Operational management is the main means of putting this into practice.

Before leaving the subject of regulatory controls and costs, it is worth exploring for a moment the current legislative position. The Environmental Protection Act introduced the concept of Best Available Technique Not Entailing Excessive Cost (BATNEEC), thus recognising the relationship of cost and environmental protection. Leaving aside the interpretation of what constitutes the best available technique and how "excessive" is determined, it is clear that the legislators were not prepared to sanction protection at any price. The current Environment Bill also requires costs to be balanced with benefits when pollution control measures are considered. In other words, the cost has to be such that the industry can stand it. While factions in Europe have made noises about the need to move to BAT pure and simple, it is reassuring to note that the EU is now referring to Economically Viable Application of Best Available Technique (EVABAT), which amounts to much the same thing as current British practice.

Environmental Management Systems

The concept of EMS, as opposed to any other management system such as quality or safety has been raised in Chapter 2. Operational management is at the core of the EMS, although chronologically certain elements of the system must precede it, while other elements follow on from it. Operations are, however, the means by which the intention is put into practice and the desired results attained, which are assessed to provide a measurement of success. Operational control can exist in isolation but unless it is included within an overall management system it is most unlikely that it will deliver satisfactory performance or give any confidence that it can play a role in crisis management.

There are now available three possible models of a management system for adoption by a company intending to embark upon the exercise; British Standard 7750, ISO 14000 series and the European Union's EMAS. It is a matter of fine judgement as to which might be the best to follow for any individual company, although as already stated, the British Standard is most widely used in this country. However, all have very similar requirements and the same purpose.

The CBI list the following eight potential benefits arising from the adoption of a systematic approach to environmental management.

- Efficiency - generally derived from better resource utilisation and waste minimisation.
- Competitive advantage - being better placed to meet changing requirements and meet exacting specifications.
- Credibility - demonstrating commitment to stake holders.
- Meeting national needs - by improving performance and reducing the need for regulation.
- Focus - bringing together uncoordinated initiatives.

These five are worthy benefits but unlike the final three they do not relate directly to crisis management.

- Compliance - ensuring that the business meets its obligations and reassuring regulators and the public.
- Risk management - able to anticipate and cope with unplanned events.
- Reduced liabilities - control of past, present and future liabilities.

Here we see that three out of the eight advantages relate directly to the topic of this book, emphasising that an EMS should be central to any crisis management system.

Environmental Management System

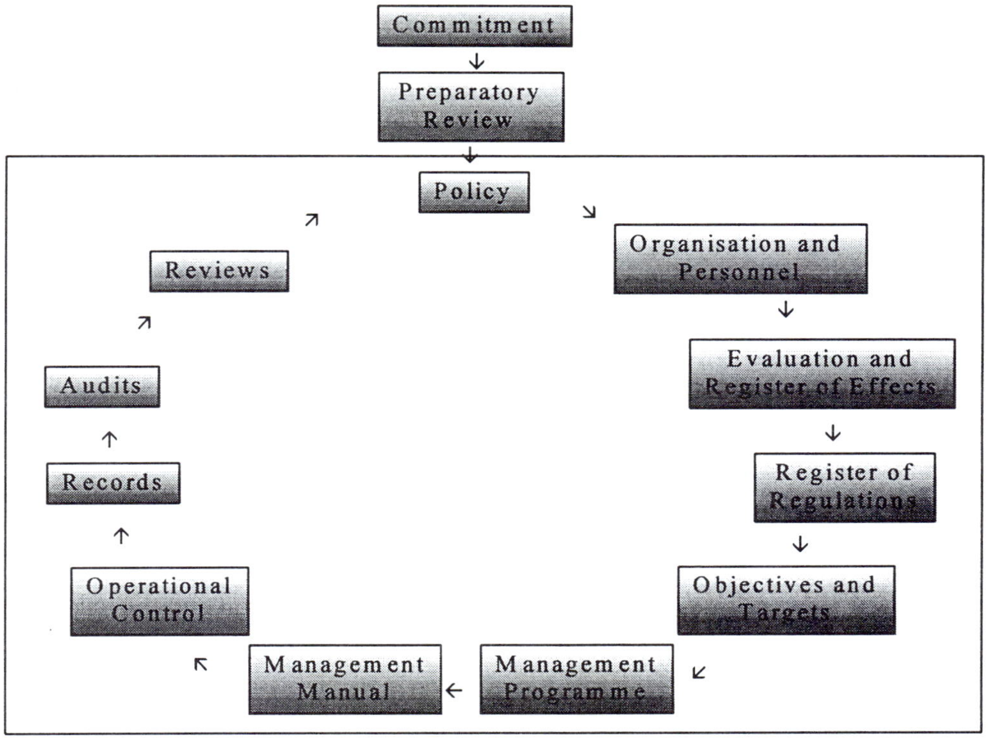

Source BS 7750

FIGURE 6

Schematic diagram of the stages in the implementation of an environmental management system

Figure 6 above, shows the management process defined by BS7750 and it is the operational control box that concerns us here. The Standard identifies various elements under the overall activity, namely co-ordinating and effectively performing control, verification, measurement and testing. In particular it details the following:

- Control including documented procedures and work instructions covering all policy aspects including procurement, monitoring and control, approval of planned processes and equipment and finally performance criteria.
- Verification, measurement and testing to demonstrate compliance with the system requirements.
- Non-compliance and corrective action to investigate and correct non-conformance and ensure that the system is adjusted to avoid further non-conformances.

While ISO 14001 and EMAS are less specific than this, their functional requirements are virtually identical. Operational control therefore covers a wide spectrum of activities such as design, modifications and developments, procurement, handling materials, working methods and training as well as emergency response plans. We shall look at each of these in turn to examine their contribution to crisis management.

The international group of companies in Bayer AG, which collectively deal with a wide range of manufacturing plants and processes together with products, all having potential environmental effects, spent 5% of its group sales (over £900m) in 1994 on environmental protection, indicating the level of benefit it sees as arising from the control of environmental risks through operational control. In particular, investment is being made in energy and waste management to bring long-term financial savings. The company considers it has reduced its energy consumption by 50% and CO_2 production by more than that amount per unit of production since the inception of the programme.

Design

Any new project concerning a product or process is required, under the BS7750 system, to have its own environmental management programme, which should, amongst other things, address risks, emergencies and liabilities. Even without an EMS it is to be hoped that no new products or processes are being introduced without at least a stand alone risk assessment or, in the case of a production facility, a HAZOP study.

The operational control at this stage is exercised by the project manager who should therefore able to identify risks and adjust the design of the process or plant to minimise them. For example the duplication of control systems, the provision of standby pumps, pressure relief values and expansion chambers, bunding, oil separators, and on site effluent treatment facilities are all means that may be considered for incorporation into the design to ensure safety and reliability.

However, the process should start at an earlier stage and might include any or all of the following:

- Site search and selection, seeking to avoid highly vulnerable neighbours such as housing, recognised wildlife sites, water courses (particularly those from which water is abstracted or which run into wildlife areas), aquifer recharge areas as well as historical hazards such as contaminated land.
- Site planning with the intention of locating the riskiest plant or processes as far as possible away from vulnerable features and including the means by which deliveries are made to the site and products or wastes are removed.
- Best available technology which would try and match the process chosen to the site characteristics, for example a "dry" process in a water protection area as opposed to a "wet" one which is more likely to lead to water contamination or a silent process where noise could disturb neighbours.
- Detailed plant design, which addresses the potential problems identified in the HAZOP study or other risk analysis and attempts to ensure that the plant or process is low risk.
- Active and passive systems to minimise and contain uncontrolled discharges. These might include self-contained site drainage system leading to a reservoir or treatment works or a "cordon sanitaire" around a storage area for volatile organic compounds.

- Fail safe systems. This is a misnomer as a failed system is a failed system but they can nevertheless be designed to fail in a low risk state.

Development and Modifications

Careful control needs to be kept over subsequent changes to the design of the plant. Any change that materially affects the site layout, process, plant or the active or passive safety systems should be subject to its own risk assessment. This should seek to identify both potential new hazards that might be created or any degradation of existing safety features.

While it might appear that this is an obvious precaution to take, it is surprising how often modifications, particularly temporary ones are, if not the root cause of a crisis, at least an important contributory factor. In particular it is important that all modifications are recorded and that management has accurate knowledge of the current state of its plant. This is an area that has itself come under new control with the Construction, Design and Management regulations, which require a site file to be maintained recording the original design, or as-built design if different, and all subsequent modifications.

Two recent cases exemplify the potential hazards in this area. Hydral Products of Warrington were fined £7,000 for the spillage of sulphuric acid that polluted a stream. The acid leaked from a flange on a tank, which was protected by a bund. Unfortunately part of the bund wall had been demolished for at least a year and had not been repaired, rendering the system useless.

Westinghouse was fined £15,000 with a further £1,000 costs for contaminating a stream at Chippenham with copper cyanide. The company correctly reported an accidental spillage and co-operated with the containment and clean-up operation. Unfortunately it did not know the drainage layout of its site so that the efforts failed to prevent the stream pollution. The court also noted that it had no training programme to alert staff to the environmental hazards posed by its process chemicals.

Maintenance

To most readers the role of plant maintenance in averting crises, through the elimination of hazards, must be self-evident. No plant or equipment can be designed and built to be maintenance free unless it is for a very limited and specific task. Maintenance requirements need to be considered at the design stage and maintenance or life cycle costs determined, along with shut down, access, spares, back-up, testing and restart facilities.

Maintenance should be preventative, that is aimed at forestalling potential hazards on a planned basis. Manufacturer's maintenance schedules should be adhered to with replacement of consumable parts before predicted failure time. The risk analysis or HAZOP study may have identified maintenance issues, allowing better planning of schedules that match the equipment to its particular use and location. For example, where two systems run in series and simultaneous failure of them both would be very hazardous, it may be better to stagger their maintenance periods so that the situation of both of them approaching the high risk failure time together is avoided. The consequence of increased down time on the plant as a whole would have to be accepted in this event. It is also important to include regular maintenance and testing of any back-up, safety and emergency response systems.

As maintenance cannot predict and eliminate all failures, maintenance systems must also be reactive, that is they must be able to cope with unplanned requirements. It is the unplanned situation that has the greatest potential to become critical. They key points in dealing with such situations are:

- A quick response that includes early identification of the problem, rapid management decision and timely reaction by the maintenance staff.
- Adequate resources both in staff numbers, competency and availability together with equipment and spares to deal with the unexpected event, even if this occurs during a heavily committed planned maintenance period. The implications are that, in order to achieve this flexibility, call-off contracts or other mechanisms may need to be in place.

Any temporary or permanent modifications to plant or processes should only be permitted after a risk analysis has been undertaken and then fully recorded, with appropriate information disseminated to all staff likely to be concerned. It is very easy to forget, for example, to carry the modification through to the training programme for new staff or to rely on "learning on the job."

Staff Training & Competency

Not withstanding the poor (by western standards) design of soviet nuclear reactors, the Chernobyl accident was entirely attributable to staff error in undertaking an unauthorised and non-standard operational process with disastrous consequences.

Operational control is exercised through staff and time and again staff are seen to be the critical element in either averting or causing crises. It is perhaps strange therefore, that in dealing with the issue of staff, two opposite, conflicting and equally erroneous views are prevalent. The first is that staff are the weakest link in the chain, will usually fail if given the opportunity and require to be backed up by mechanical, automatic systems. The second is that staff are dependable automata, who will always carry out their appointed task and in a crisis or potential crisis, and can be relied upon to take the correct decision.

The truth of course lies somewhere between these extremes. Analysis of past failures suggest that the common causes of human error are:

- Inadequate information or ignorance of the job and its potential consequences;
- Poor communications between management and either worker or shifts at handover;
- Lapse of attention particularly by skilled workers who are attempting to achieve too much;
- Mistakes made in the belief that the wrong action taken will have the intended effect;
- Mis-perception, which often occurs under stressful situations;
- Wrong priorities applied by the individual when faced with conflicting demands; and,
- Wilfulness or poor discipline such that systems are overridden, circumvented or ignored for whatever reason.

It is interesting to reflect on how many of these causes can be identified as not only contributing to a crisis but actually exacerbating a crisis once one has occurred. To guard against these dangers operational control needs to consider the following.

Command Structure - This is not necessarily the same as the management structure. The line of command relates to the authority for taking immediate operational decisions on site and the transfer of that authority in the absence of any particular individual.

Numbers and Resources - Few companies will be in a position to employ more than the minimum number and grade of staff necessary to undertake the task of managing environmental issues. This means resourcing down to the level that is appropriate to the situation when all is going according to plan. Consequently all unplanned staff events, such as sickness, and even planned events such as holidays and training, can potentially leave the company open to risk. How cover is maintained, in terms of numbers and competency, must therefore be addressed. Various options exist such as the use of agency staff, overtime, flexible working and call-off contracts with outside firms, but none are without problems and the choice needs to match the particular situation.

Competency - In this context I use competency in terms of the ability to do a specific job that arises from sufficient training, qualifications and experience. This may seem obvious enough but it is not uncommon to find, for example, in an engineering based industry, that an engineer is made responsible for environmental matters. Whether this individual needs any additional training or experience is often over looked and the option of employing an environmental professional may never be considered. Firms need to identify the environmental aspects of any given job and the competency required to adequately discharge the duties. The NCVQ approach may be a good starting point where no job specific requirements already exist.

Job Design - Ergonomic design of the tasks, the working environment or the product to take account of human capabilities and weaknesses, can make human error less likely. However, it is depressingly common to find that the operational design intent is not followed by those operating or using the design in practice. For example, in an attempt to overcome problems associated with deliveries, one company ensured that delivery lorries and tank fillers were gradually fitted with specific and non-interchangeable couplings so that, for example, oil could only be delivered to an oil tank filler. However, a later audit revealed that, in order to be able to deal with "non standard" systems, delivery drivers had provided themselves with a range of adapters so that the delivery pipe could be mated to any filler!

Motivation - Even with sufficient staff of demonstrably competency and equipped with sufficient resources, adequate performance can not be guaranteed unless they are motivated to do the job well. Protection of the environment, through, amongst other things, the avoidance of crises, needs to be appreciated by staff as important for both business and ethical reasons. Staff need to identify with the environmental achievements of their company. Motivation is not the same as financial reward, although the two may be linked. The relationship to be aimed for should be that of the provision of training and essential resources to increase awareness and motivation. That heightened motivation should then lead to better performance, which can be used to finance both the training and resource costs and appropriate rewards.

The Engineering Council identify the major considerations in the design of the job as:

a) Identification and analysis of the critical tasks expected of individuals and appraisal of likely errors;

b) Evaluation of required operator decision making and the optimum balance between the human and automatic contributions to safe actions;

c) Application of ergonomic principles to the design of man-machine interfaces;

d) Design and presentation of procedures and operating instructions;

e) Organisation and control of the working environment, including the workspace, access for maintenance, lighting, noise and thermal conditions;

f) Provision of correct tools, equipment and protective clothing;

g) Scheduling of work patterns including shift organisation, control of fatigue and stress and arrangements for emergency operations.

Attention to these factors will lead to the conclusion that staff performance is largely in the hands of the employer but is seldom ever likely to be wholly dependable.

But by the same token performance is not going to be so unreliable that staff have to be backed up by automatic systems. In fact the latter attitude may be self fulfilling in that staff may come to rely on the automatic systems and see themselves as being superfluous. In the event of a crisis arising they are then unlikely to be able to respond effectively. Accordingly, the staff element of operational control needs careful integration with the other operational control elements and recognition needs to be given to the fact that performance can substantially modify, for either the better or the worse, their effectiveness.

Lanstar of Manchester recently discovered the fallibility of systems dependant on their staff. The NRA made a routine visit to the company and, when the visiting officer was not met as usual by the firm's Safety and Compliance Officer, he went unaccompanied to the site outfall. Here he found the lady he was expecting to meet attempting to close the outfall to the Manchester Ship Canal. One hour previously a container of hydrochloric acid and titanium oxide had been punctured and some 500kg had been discharged to the site's drainage system and to a catchment pit. However, due to the absence of the Engineering Services Manager, who was attending a dental appointment, the normal procedure of isolating the catchment pit from the outfall had not been followed. The dental appointment cost the company about £8,500 in fines and costs, although, to be fair, it was not the appointment but the system itself that was the root cause of the problem. Lack of consideration of the staff element, in particular the chain of command, and procedures, possibly compounded by a lack of training, were all contributory factors here. In passing the wisdom of having a discharge system, whose failure mode was in the "open" state, might also be questioned.

Procedures and Work Instructions

British Standard 7750 suggests that procedures, which are considered to be high level, company-wide documents, and work instructions, lower level, often site specific instructions, are needed for any activity or process that could have significant environmental effect. Significant effects are not the same as potential crises. The former may arise for example, through long-term purchasing policy whereas we are more concerned here with, say, an accidental discharge. Again, implementation of a recognised management systems should ensure the control of those issues likely to give rise to crises. However, for a company that can not commit itself to an EMS or that needs immediate action to minimise the risk of a crisis, procedures or instructions to address operational control can be implemented on a stand alone basis.

The process for establishing procedures should be structured and may beneficially be considered as passing through the following stages:

Establish the need which may arise from the policy, legislative or regulatory requirement or risk control. This will identify the purpose and scope of the proposed procedures.

Define and document the elements and flow of the process that need to be controlled.

Drafting for which the identification of an appropriate author, who is competent and ideally a stake-holder in its implementation, is vital. The author will need to be clear in his or her own mind of:

- the purpose of the procedure,
- what activity is to be defined and controlled, - who will implement it,
- how it will be done,
- when and where it will be done,
- how it will be verified.

Trialling a draft procedure to identify its acceptability, practicality, effectiveness and any potential problems.

Revision through reworking the draft in the light of the trial results.

Issue of the procedure to relevant staff including any training and equipment that is consequential upon it, so that it is both implemented and effective.

Verification mechanisms that ensure its effectiveness is proven and auditable.

Review on a regular but infrequent basis.

Work Instructions, while appearing to be of more local, and therefore of perhaps lesser importance, should nevertheless evolve by a similar process. The fact that they may relate directly to the operation of a piece of equipment or the running of a process may mean that they are easier to conceptualise than a procedure. However, it is still important that the purpose of the instruction, which will often be the safe operation of the process in both normal and abnormal operating conditions including start-up, shut-down and maintenance, is firmly kept in mind. It may be helpful for the author to approach the exercise from the viewpoint of an inexperienced person or one with limited competence in the matter, who is charged with the control of the activity.

Consideration must be given to monitoring the procedure in order to verify its effectiveness. This may itself require a separate procedure together with the establishment of checklists, record forms and equipment registers along with their calibration and maintenance. The reporting of incidents and action arising from them must also be covered and hopefully the majority will be relatively small, if not trivial.

Nevertheless, reporting of all incidents is important for several reasons. It raises matters to the attention of management, it can provide useful data on the working of the system and it may provide early warning of equipment failure or serious non-conformance with procedures. Effective reporting is often difficult to establish, not least because employees see it as a mechanism by which management is made aware of the employee's own shortcomings. The only way around this is the establishment of a "no blame" system of reporting, investigation and corrective action. The question to ask is "how can we fix it?" not "whose fault is it?"

Line management must be fully conversant with procedures and must be seen to ensure not only that they are followed, but to conform to the requirements themselves. It is potentially dangerous to have a written procedure but to let, by custom and practice, a different method of working evolve. In such cases either the procedure must be reviewed or discipline enforced. The advantage of an environmental management system is that it should prevent such a situation developing.

Management Commitment

Ultimately operational control comes down to the commitment of management to ensure that day-to-day business activities are defined, controlled and monitored. There is a natural and understandable tendency to concentrate on the project initiation phase through to commissioning and then, when that has been successfully achieved, for attention to pass to other issues. Ironically this shift in attention occurs at the same moment as the most significant risks begin to be taken. After all there is little risk other than those arising through the construction process until the process itself is actually up and operating. There are a number of reasons why this attention lapse can happen, such as:

- Loss of novelty value and interest,
- Need to save costs,
- Transfer of responsibility to others who have been involved in the project to a lesser extent,
- Fading of memory of why things were done in a particular way,
- Desire to leave something that is working, well alone and concentrate on problem solving.

Any or all of these can lead to a lowered management commitment and greater risk as a result. There are also a number of factors which can focus the attention of management and might, as a result, be encouraged in order to maintain commitment. These fall into two groups, those over which management have little control and those that are largely in the remit of management to control. In the first group are the legislation and regulatory requirements that must be met. In the second group are:

- Seeking continuous improvement, perhaps driven by BS 7750, or similar,
- Benchmarking against competitors,
- Cost saving through resource and waste minimisation,
- Communication with stake holders, who have higher expectations of performance than management is currently delivering.

The converse of management commitment was exposed in the report into the Piper Alpha oil rig disaster. This found significant flaws in the emergency procedures, which stemmed from senior management being too easily satisfied with existing procedures, and for failing to provide training in the face of a known problem. The report concluded that managers had adopted a superficial response when safety issues had been raised. Do not let similar critisims apply to your management of the environment.

The Oil Delivery Audit

BP and Alcan Rolled Products were each fined £15,000 in March 1995 after 550 gallons of oil polluted the River Ebbw. The oil delivery took place in darkness in the early hours of the morning when there was no site supervision. Heavy fuel oil was pumped into a tank that was under repair and, unfortunately , so was its bund. BP knew of the repairs but had not briefed the driver. Alcan had failed to put any warning notices on the tank or to lock the tank or the cap on the delivery pipe. The clean up costs to the two companies was in excess of £60,000.

Oil deliveries, like many other materials, are routine, not only in industry but also in domestic situations. Yet it is surprising how often routine deliveries go wrong. It is worth considering the reasons for this and the consequences to see what lessons can be learnt.

Current legislation imparts strict liability for environmental failures in most instances, so that chairmen, directors and senior managers can find themselves held personally responsible as well as the corporate body. Delivery arrangements need to be high in management consciousness, carefully controlled and regularly audited. If an audit has not been carried out recently it is perhaps a good idea to consider the problem from this perspective. Outlined below is the audit process for looking at oil delivery and storage, but the process could readily be adapted to consider any aspect of the site or operation.

The audit should start with a questionnaire to be completed by site managers and operational staff to assist with identifying the site features that will need investigation. The preparatory phase of the audit should establish the basic facts concerning:

- Why deliveries are required at that location
- Equipment
- Operating procedures
- Management system.

As in any audit it is worth spending a considerable time on the preparation phase to ensure that the auditor is thoroughly conversant with the process and that the right issues are followed up on site.

The main issues in terms of equipment will be the means of containing spillages and the site drainage system. Checking of the following points will establish the extent and state of the infrastructure:

- Identification of tanks and their filler pipes
- Impervious surfaces and barriers to contain spills
- Layout and maintenance of drains
- Number, capacity and maintenance of oil/water separators or oil interceptors
- Vehicle parking and washing facilities, which may need separate drainage systems to avoid detergents and emulsifiers passing into interceptors and reducing their effectiveness
- Volume of bunding which should be 110% of the largest tank contained within
- Means of disposal of rain water from the bund
- Accurate and working gauges on tanks
- Stock records to reconcile inputs to usage. Any inconsistency could be due to hidden leakage, which is particularly serious for underground tanks.

Operational issues will revolve around the documented procedures and the communications used on site to instruct and inform staff. The site audit will need to concentrate on:

- Condition and labelling of inlets including lockable caps and pipework
- Emergency stop-cocks, their identification and operating instructions
- Presence of drip-trays and contents
- Means of storing and disposing of waste oil
- Emergency equipment
- The actual working practices of staff particularly the relationship and responsibilities of delivery drivers and site staff.

Finally some thought should be given to pollution control in the event of a fire, especially the containment and disposal of fire fighting water and foam.

On completion of the audit a report must be prepared, which covers the factual situation, identifies deficiencies and the corrective actions that are necessary. A timetable for their implementation and a re-inspection of the site should be agreed with management.

The areas of concern for the auditors, that are outlined above, indicate that there are many potential causes of oil delivery problems. Management must realise that simply providing the means will not ensure that it is used correctly. As outside parties are also involved in deliveries then, to control risks adequately, requires that the system must also control the actions of other staff temporarily on the site. In fact it would be advisable to invite the relavant regulatory body to take part in the audit. In this instance the NRA would be appropriate and it would probably be in a position to provide practical advice, for it is keen to work with industry whenever possible. Such co-operation is likey to stand the company in good stead if disaster does strike.

Auditing plays a vital role in operational management. Without it management can have no confidence that control is being exercised or that it is relevant and up to date. It is important that trained auditors are used, who may perfectly satisfactorily be internal to the company although they should not be directly concerned with the function being audited, and that a regular programme is set up. It is also important that the auditees see the process as a helpful one that assists them in improving performance and not a means of apportioning blame. Auditing is a specialist topic in itself and the reader is referred to "Environmental Auditing: A Guide to Best Practice in the UK and Europe" in this series for an in-depth treatment.

PART 4 CRISIS MANAGEMENT

7. *Planning for the Emergency*

In spite of taking the precautions that we have considered in the earlier chapters there will come at some point the unplanned incident that is, or has the potential to become an environmental crisis. This chapter looks at what can be done to prepare for that crisis and what can be done to control and minimise its effects.

The circumstances in which a crisis arises, whether it is an accident, some other incident or emergency, differ widely so making planning more difficult and requiring flexibility in both plans and personnel. The response to an oil spill must clearly be very different to the response to undesirable environmental side effects of a product that has been widely marketed for five years. The risk analysis will provide pointers as to the likely nature, location and frequency of any incidents but can do no more. If it could, then almost by definition we would have no crises as they would be, or should be, foreseen and eliminated. As a result emergency planning involves a significant commitment in time and effort and will inevitably be planning for many events that never happen. This cost must be accepted as part of the overall costs of crisis management.

Environmental crises may arise that concern only the environment and have no other implications but this will not always be the case. A fire in a petrochemical plant, for example, threatens to cause uncontrolled discharges to air and water. It may also directly threaten the safety of employees, neighbours and the emergency services. Additionally it may also leave a long-term land contamination issue. Further, if we accept the financial element of the definition of an environmental crisis, then the financial threat arising from loss of production and loss of productive capacity must also be recognised. Thus a crisis may have many dimensions and environmental crisis planning needs to be integrated with other emergency plans in order to deal with these wider issues. For many companies this will mean that planning for the environmental crisis will not be the creation of a new system by some unfamiliar process. Rather it will be part of the continuous review and extension of existing plans to cope, in this case, with the environment, in the light of the new risks and opportunities that are emerging all the time. As Mark Scoggins of solicitors Davis Arnold Cooper says: "You have to have, first of all, a plan and secondly you have to make sure it works reasonably well. A chairman of a public inquiry is going to be extremely interested not just in how well you performed in preventing the incident but how well you performed in managing it."

It may in practice be found that the extension or integration of the plans is a relatively straight forward task. Most actions to control "conventional" crises also serve a similar purpose in environmental crises and the same emergency services are involved in both cases, even if they perform different roles according to the situation. Netherless, it is the company's management that is still likely to face a challenge when trying to integrate and balance the need for continuity with the sometimes conflicting demands of an evolving crisis.

The hypothetical fire above will illustrate the difficulty. The plant might be expected to have a through put value of £1 million a day representing an operating profit of £100,000/day. Pressure will be on management to minimise the effects and costs of the fire and, after ensuring personal safety and extinguishing the fire, one might consider that restarting production within 24 hours would represent a considerable achievement, containing loss of production profits to £100,000.

However, if this has been achieved by ignoring the environment, the situation looks somewhat different. Consider that polluted fire fighting water is allowed to run off the site and into a controlled water course. Damage occurs to the local ecology and a water abstraction point two miles downstream has to be closed. Water supplies to 250,000 people are disrupted. Meanwhile, combustion products are released as well as dust arising from repairs and drifts off site, contaminating the three neighbouring farms, which have to cease milk production for several weeks. Prosecution by the NRA, fines, damages through civil actions and clean up costs amount to over £1 million. Loss of good will with regulatory bodies and neighbours and damage to reputation is not costed directly, but is substantial. Did local management make the right decision in concentrating on resuming production as soon as possible? Bear in mind that local management probably had little understanding of the environmental issues and were no doubt going to have to account for the fire in the first place. It is very easy to take decisions at the height of the crisis, which go against the principle of integration of emergency plans, if indeed the environment is actually covered by a plan in the first place. It is also easy to make an over hasty response. Washing a spillage away into the drains may turn an incident into a crisis.

The Crisis Management System

Crises are individual but, while no two are the same, all tend to have elements in common. They are usually preceded by a series of seemingly trivial and unrelated incidents, some of which may prove to be embarrassing when they come to light, that are overlooked if not actively ignored. The typical crisis will come as a surprise and will occur at awkward times such as Friday afternoon and bank holidays. Key staff are absent and equipment temporarily under repair. Management will first be made aware of the crisis by a phone call from the media, whereupon there will be constant pressure for answers and action from community groups, customers, investors and the rest of the media. The regulatory authorities will expect immediate action and the emergency services will arrive and work to their own plans. Site management will be overwhelmed and the situation will be out of control.

The one thing that the company management needs but does not have is hard facts and it has to operate on hearsay and guestimates. However, the agenda is soon set by external "experts" whose speculation comes to be accepted as fact and media scrutiny soon takes on a judgmental aspect in so far as management performance is concerned. Not surprisingly, management responds by adopting a defensive posture and confining itself to reacting to criticism and attempting to limit damage. So runs the "typical" crisis but clearly no one can consider this a satisfactory situation. What can be done to avoid it?

We will be looking in detail at the emergency plan itself and the means of dealing with the media in more detail in Chapter 9. Management action, however, goes beyond these and needs to concentrate on actively managing the crisis. Only by vigorous, determined and purposeful action can the initiative be gained and all aspects of a crisis brought under control.

Three Basic Requirements

There are three basic requirements for successful crisis management. The first thing is to define the real issues involved in the crisis - what has happened, what are the likely consequences and what needs to be done? Secondly, management needs to be free to concentrate on the crisis, implying that either a dedicated manager should be appointed or the existing management must be able to delegate all day-to-day management issues and so leave themselves clear. Thirdly, a crisis management centre needs to be set up to implement the operational plan and all information and decisions directed through it. Only by means of satisfying these basic requirements can the initiative be gained and the crisis be brought under control. Fortunately, these do not need to be created afresh for each crisis. One plan can cover many potential crises.

The Emergency Plan

The means by which the above requirements are fulfilled represent the core strategy of emergency planning. It goes without saying that an emergency plan needs to be in existence and understood if it is to be effective, and to ensure that this is the case, its must be put in hand well before any crisis. The crisis itself is no time to be testing and refining the plan, let alone formulating it.

We speak of "the plan" but in reality, there may need to be a series of complementary and interlocking elements covering all the significant risks on the site or sites, and possibly off site risks as well when these can be shown to be the company's responsibility. Furthermore, these plans need to link with safety and other similar plans and to be co-ordinated with the emergency services.

The emergency plan should cover the following operational issues:

- Who will take responsibility for its implementation.
- What procedures will be followed.
- Known risks and particularly sensitive environmental resources.
- Linkage with the emergency services.
- Any linkage to other plans.
- Equipment and manpower resources.
- Training and testing.
- Incident reporting.
- Recovery and restart process.

A template of a typical emergency plan which covers all these issues is attached at Appendix D. In the following paragraphs each issue is discussed individually, which should enable the reader to modify and complete the template to cover their own site or circumstance in order to fulfil the three basic requirements.

Who Will Take Responsibility

A clear chain of command must be established so that somebody is identified as being in charge, with deputies or substitutes identified in the event of key personnel being unavailable (leave, sickness etc.). The authority vested in each member of the crisis team should be defined in order to enable rapid decisions and executive action to be taken. In an emergency it may be necessary to ensure that the man on the spot has considerably enhanced delegated power to enable him to act effectively. The means of alerting the team (telephone numbers, pagers etc.) must be defined. In addition, all other parties, who may be involved, need to be identified together with their contact numbers, e.g.

- emergency services
- regulatory bodies
- local residents/community group/local authority
- external resources that can be used (specialist contractors).

What Procedures will be Followed

Clearly these will vary from site to site but should cover as a minimum:

- Alerting/alarm to warn persons in immediate danger and key players in the emergency response.
- Evacuation and dealing with casualties.
- Control and physical security of the site.
- Identification of the facts and key issues.
- Containment of the immediate problem.
- Making the rest of site safe and possibly resuming operations on it, if practicable.
- Treatment, reduction and removal of contaminants, damaged equipment etc.
- Use of emergency equipment.
- Incident reporting.
- Public relations and information control.
- Repair, Recovery and Restart.

This list is not necessarily exhaustive and may well need to be extended in the light of testing (see below) or of experience gained in an earlier crisis.

A simple executive summary, possibly in the form of a flow diagram and decision tree can act as a quick reference for the action to be taken and the inter-relationship of the procedures, and is usually placed at the head of the documentation. The more that can be covered as a prepared procedure, the more time management can devote to dealing with the specific crisis issues.

Risks and Sensitive Resources

The plan should follow from, and relate directly to the risk analysis. The main risks should be identified in the plan and also the sensitive environmental resources that could be affected. Knowledge of these will ensure that the plan addresses the likely hazards and will materially assist the crisis management team to identify the key issues and react accordingly. In addition site details such as pipe runs, storage tanks and drainage plans should be included as easily accessible appendices.

Links with Emergency Services

Under the heading of emergency services should be included regulatory authorities, who can and indeed will, take action in a crisis to protect those elements of the environment for which they are responsible. Again contact numbers, key individuals, information and reporting requirements need to be identified. The assistance that can be rendered by them in terms of technical expertise and equipment can usefully be summarised. The requirements of the police, who would be expected to take operational control on site in the event of a crisis, should be considered. While they will have mobile operations rooms and communications, it may be preferable if these facilities can be provided from the site's own resources. Likewise vehicle routes, evacuation procedures, holding areas and possible casualty collection arrangements should be covered. (See chapter 8 for further consideration of the emergency services).

Linkage with Other Plans

Besides the plans of the emergency services, the company itself may well have existing plans covering the issues set out below and the environmental emergency plan needs to be checked to ensure it dovetails with these and defines the relationships and priorities.

- Health & Safety,
- Quality,
- Alternative supply, production and distribution,
- IT and communications back-up,
- Customer care,
- Public relations
- Loss control and insurance

Equipment & Manpower Resources

The location, specification, capabilities and quantity of emergency equipment held on site, or off site if it is readily available for use in an emergency, must be recorded. Likewise the manpower that can be mobilised either to operate the equipment or assist in more general ways should be identified together with the means of call out and expected speed of response. Where it is not realistic to provide dedicated personnel and equipment on permanent standby on site, it may be necessary to arrange call-off contracts with specialist companies for the supply of these resources on a 24 hours basis. Again phone numbers and response times must be noted. However, be aware that in a general or very large scale crisis even the specialists may be over-committed and unable to fulfil their contractural obligations.

Training and Testing

Personnel need to be trained and both equipment and systems tested. The emergency plan should define the training required of the crisis management team itself and other staff, and the training delivered together with the competency achieved. Exercises or refresher training or both will probably be required to keep people up to date and fully competent. Clearly all key staff need to be totally familiar with the emergency plan and it may be worth considering the issue of a pocket aide memoir to all staff covering the vital response to ensure consistent and effective action.

Equipment needs to be regularly inspected and tested and a schedule established for defining the programme and recording the results. Likewise the emergency response system needs to be regularly tested, either by table top or computer simulations or, more rarely, on the ground, to ensure its proper functioning and that personnel are practised in their roles.

Staff must be provided with and trained in the use of personal protective equipment. Sufficient stocks ought to be held to cover high wastage rates during an emergency and, if appropriate, thought given to decontamination facilities and contaminated waste disposal. Remember that the testing of emergency plans is a requirement of BS 7750.

Incident Reporting

All incidents that occur, whether or not they could be classed as an emergency, need to be logged if they affect the environment or relate to a matter covered by, for example, discharge consents. The log should record the action taken as well as details of the location, magnitude, cause, consequences and the person responsible. It is an important tool for investigation into the possible cause of an emergency.

The log can be used to build up a better picture of the risks on the site and the reaction to them, allowing the emergency plan to be adjusted accordingly and action to be taken to reduce or eliminate risks. The log will also be a key document when dealing with regulatory authorities, who will interest themselves both before and after a major incident with the safety of the plant.

Recovery Process

It is clearly important for a business to recover as quickly as possible from a crisis and indeed until recovery is complete the crisis can be said to still exist, albeit in a diminished form. The emergency plan should either cover the process or better, it should integrate with a separate recovery plan. At site level this will need to deal with:

- Temporary alternative production facilities.
- Dealing with stake holders.
- Public relations.
- Repair and replacement process.
- Restart process.
- Modifications and lessons learnt.

Different requirements exist at company level and are considered in chapter 9.

In conclusion, there may appear to be a daunting amount of work required to put together a plan and maintain it in an effective state. The questions must be asked, is it worth it and can it be reduced?

The answer to the first question depends upon the results of the risk analysis. If the business does face significant environmental risks then a plan is a necessity and that plan must be comprehensive. Remember that once the initial investment in time and effort has been made to prepare the plan, then review and updating is not that onerous. An effective plan and trained staff are likely to assist with developing good relations with regulatory authorities and may save time in both negotiations and monitoring inspections.

The plan needs to be sufficiently comprehensive to be able to deal with all significant risks. If it is only a partial plan, you may be certain that the event that has not been covered will be the one to occur, leaving you to learn on the job and rendering the effort you have put into it already, largely wasted. Chapter 10 takes the reader through the checklist to help identify risks and the template for drawing up an emergency plan, which will ensure that management can rapidly:

- find out what is happening
- gain the initiative
- maintain the initiative and manage the crisis to a successful conclusion.

8. *The Emergency Services*

Your environmental crisis does not occur in isolation from everyone else. By definition it affects the environment which is wider than just the company and its property. In that wider environment it must take its place within a spectrum of potential crises for which there already exists, to a greater or lesser extent, the means to deal with them. Some of those means, notably the regulatory bodies such as NRA and their abatement and clean-up powers, we have already touched on. The more usual emergency services, police, fire and ambulance and the civil system for coping with emergencies must now be considered, so that company and civil emergency plans can be effectively integrated.

The Civil System

At local level, responsibility for the co-ordination of different services within a unified framework rests with local authorities. Various different approaches exist but the one that is generally preferred is to delegate responsibility to the police to lead a group consisting of:

- Chief officers of the emergency services,
- Local representative of the NHS,
- Local authority chief executives,

who are charged with planning, formulating procedures and training. Such a group may form more junior working parties to deal with specific issues.

The emergency plan will seek to co-ordinate the actions of, not only the parties on the planning group, but also the voluntary sector, the NHS and commercial and industrial interests where these are judged to be a significant local hazard. At the very least, nuclear installations and those covered by the CIMAH regulations, which are looked at in more detail later, must be considered. The plan should pay particular attention to:

- Dealing with the media.
- Communications, both in terms of sufficiency and back-up systems.
- The best use of the voluntary sector and the resources available.
- Co-ordination with neighbouring areas.
- Ability to sustain action over an extended period.
- Training and exercises.

The latter is potentially a substantial commitment but much can be done to train individuals and test procedures with table top exercises and computer simulations. Certain exercises must be undertaken according to statute and these include the nuclear and CIMAH installations, HM Coastguard and the railways. There is no specific requirement for plans to deal with environmental emergencies and practice may differ substantially at local level. It would certainly be good practice to contact the local authority emergency planning officer, whenever significant risks are identified, both to seek information on the existing plan and to explore the opportunities for integrating the firm's own emergency plan with it.

The Police

The police will co-ordinate activity at the crisis site, assuming that the effects extend beyond the plant boundary and are significant. This co-ordination may be exercised through the establishment of an Incident Control Post, which may beneficially be co-located with the crisis management team's own facility. They will process casualty information and deal with the identification and removal of the dead.

They are required to treat the emergency site as a scene of a crime and they also have certain responsibilities for investigation of the cause of death or for facilitating the HSE and others in their investigations. Accordingly, expect close co-operation when dealing with the immediate emergency but an "arms length" relationship in respect of their investigative responsibilities.

The Fire Service

The fire brigade has a wide range of responsibilities and capabilities when it comes to environmental crises arising from industrial accidents. This will extend beyond fire fighting to encompass rescue and dealing with chemical spillage's so that the site can be rendered safe. However, the service will not deal with medium or long-term contamination, the responsibility for which remains with the company under the control of the appropriate regulatory authority.

The Ambulance Service

Unless there are casualties or suspected casualties requiring observation, the ambulance service's role in the environmental crisis is likely to be very limited. Where there are injuries or death then the service's priority will be first aid, to stabilise the casualty's condition, and then evacuation to a suitable receiving hospital.

The Local Authority

The local authority will support the emergency services as appropriate with, for example, the provision of heavy engineering or other specialised equipment. It will also deal with the care and welfare of persons that have to be evacuated and once the recovery phase is reached, take responsibility for restoring the environment.

The Response to a Crisis

The Government has set objectives to which all organisations and agencies should be working in the event of a disaster. These are:

a) to save life;
b) to prevent escalation of the disaster;
c) to relieve suffering;
d) to safeguard the environment;
e) to protect property;
f) to facilitate criminal investigation and judicial, public, technical, or other inquiries;
g) to inform the public;
h) to promote self help and recovery; and
i) to restore normality as soon as possible.

It should be noted that protection of the environment is a relatively high priority reflecting the importance that is now attached to environmental issues.

The need for accurate information is seen as vital in ensuring that an appropriate and co-ordinated response is made. To that end the first representatives of the emergency services to arrive on site will be expected to assess the situation rather than become immediately involved in dealing practically with the crisis. Thus the more information that can be supplied by the company's own crisis management team, the faster the emergency services can be deployed to deal with it.

The measures that the co-ordinating service responsible for setting up the incident post, probably the police, will be considering will be:

- Making best use of resources and expertise by assigning responsibilities for specific functions to the most appropriate organisation.
- Securing the affected area by setting up a cordon and restricting entry, controlled by the incident post. This allows control of the numbers of persons involved on site and accounting for them should the problem escalate.
- Casualty collection and accounting, evacuation of survivors and threatened persons.
- Rendezvous point for emergency services, traffic control, marshalling and route clearance.
- Temporary mortuary and identification of the dead.
- Control of the media through provision of a liaison point.
- Depending of the scale of the emergency, the establishment of an outer cordon and air traffic control may be necessary.

Many of the measures clearly parallel the concerns of the company's crisis management team, emphasising both the need for pre-planning co-ordination and close working on site.
The final concern will relate to the collection of evidence to assist in any inquiry or judicial process and the keeping of an incident log, both to assist with immediate control issues and with lessons learnt.

At first sight, few disasters, which involve the emergency services, arise from an environmental crisis. Yet closer examination shows that few disasters are without environmental side effects, so that the issue is a very real one. For any company but particularly the SME, the lesson is that all crisis plans need to be interlinked as a small fire, for example, can rapidly turn into a major environmental crisis, while almost any fire will have environmental consequences.

To this end, it is worth ensuring that the environmental emergency plan considers risks and consequences that could arise that are not strictly environmental in nature. These include:

- Physical security of the site, control of visitors, office security, intruder alarms and sensors and CCTV. These measures are particularly pertinent for companies that might be subject to terrorist attack, where a deliberate act of sabotage could be intended to initiate an environmental crisis.
- Fire risks, equipment, training and emergency plan and procedures.
- Site services namely normal and back-up supplies of electricity, water, gas and telephone and the treatment and disposal of wastes.
- Physical condition of buildings, equipment and plant, maintenance procedures and means of access and escape.
- Health & safety organisation and responsibility, training, equipment and monitoring of exposures and health.
- Information technology, its vulnerability to disruption, security of data, continuing operations if temporally unavailable or permanently lost.
- Insurance cover, risks either passed on or carried by the company, levels, conditions and exclusions.

All these issues impinge on the environmental emergency management plan and the effective interface of that plan with other emergency plans or provisions. While it may be possible to consider and plan for these issues as a paper exercise, testing, that is using staff in the roles that they must play in the emergency will be necessary if gaps and weaknesses are to be eliminated.

CIMAH - The Special Case

Reference has already been made to the Control of Industrial Major Accident Hazards Regulations (CIMAH) 1984, subsequently amended in 1988, 1990 and again in 1994. It is assumed that no company that is potentially or actually subject to these regulations is unaware of them. However, for many SME's particularly those operating in other business sectors the regulations may be unknown. The significance of them is that they are the closest approach legislation makes to an environmental crisis management system. As a result it is worth considering CIMAH and comparing it with the suggestions in the foregoing chapters.

A number of well publicised disasters in the past quarter century, such Flixborough, Seveso, Bophal and Piper Alpha, lead to a growing awareness of the potential hazards posed by modern industrial concerns, which store and process relatively large quantities of dangerous substances. As a result both Britain and the EC have brought in regulations to control these operations. In Britain the system is based upon:

- identification
- location
- risk assessment, particularly in respect of neighbours

The system builds upon earlier regulations that required the storage and handling of large quantities of certain specified substances be notified to the Health & Safety Executive. The CIMAH Regulations apply to all industrial activities such as processing chemicals, refining, incineration and gas fuel production, including storage of any substance defined in the schedules. The HSE requires the operator to provide a "safety case" that demonstrates how the process will be operated safely. The safety case must include a formal and structured risk assessment and an explanation both as to how the risks will be controlled and the consequences of any accident, affecting persons or the environment, minimised. It should also detail staff responsibilities, training and the equipment provided for their protection. The safety case is an abstract from a much larger body of information about the plant and its management, which enables the HSE to judge if the risks are properly identified and managed. Responsibility for the safe working of the plant still remains with the operator. To discharge this responsibility, the operator must prepare site emergency plans, in consultation with the HSE and the emergency services. Staff responsibilities (in the event of an emergency occurring) must be defined, and all staff affected by the plan must be informed of its provisions.

The local authority is made responsible for preparing an off-site emergency plan. It must consult with the operator, who is obliged to provide information on the possible off-site effects in the event of an accident, and also with the HSE and other relevant agencies. Both the on and off-site plans must be kept up to date and take account of all material changes in the plant and process.

Neighbours and members of the public living in the vicinity of the plant must be notified of the risks arising from the plant. Again the local authority is responsible for disseminating this information. It is also likely to want to make the relevant parts of the off-site emergency plan public in order to improve co-operation and effectiveness. Matters covered should include the

type of activities on the site, the common names of harmful substances likely to be released, the warning or alarm system together with basic safety measures to be followed.
The reader will recognise many elements in this legislation that mirrors the measures that have been explored in the preceding chapters, though it may at first sight be more concerned with health and safety. However, the "Seveso" Directive is explicit that the rules are equally applicable to the protection of the environment from the risk and consequences of a major accident. This is defined as causing

"permanent or long-term damage to particular unique, rare or otherwise valued components of the man made or natural environment or if there is widespread environmental loss or damage".

The Directive also includes advice about desirability of undertaking baseline surveys, monitoring and investigation of the ecotoxicological characteristics of any substances that might be released.

Currently the EC is considering the Control of Major Accident Hazards (COMAH), regulations that closely mirror the British CIMAH regulations. One significant difference is that public consultation will have to take the place on emergency plans. Companies must therefore expect increasing public interest in, and scrutiny of, their operation and the risks that they pose, together with the measures that they are taking to control them. The situation will not, therefore, be getting easier. Now is not too early to start planning.

9. *Public Relations and Reputation Protection*

These two issues are intimately bound up with crisis management. Of the two, public relations can be considered as the tactical level and in the brief description of how most crises develop, we have seen that the media can often both define and drive a crisis. Reputation is a longer term, strategic issue. It may survive a crisis but will almost certainly be damaged. It will not survive a series of mismanaged crises.

It is worth pausing to reflect that relations with the public can actually precipitate a crisis. Deeside Aluminium of Wrexham has found itself in receipt of HMIP improvement notices. The company was producing excessive smoke, which arose from oil or grease on aluminium scrap that was being processed. However, the significant point is that HMIP's attention was drawn to the problem by local residents, who had made video recordings after becoming concerned about the company's general environmental performance.

Public Relations in Crisis Management

One of the primary objectives of the crisis management team should be to gain the initiative and to do this it is necessary to control information flow and manage the media. The early actions of the team in collating the facts of the situation, defining the critical issues and setting up a control centre through which all communications are routed are the necessary elements to ensure crisis can be successfully managed.

It is important to understand the media and their motives if they are to be managed successfully. Their primary interest is to provide information to the public and to do this effectively they must first catch the public's attention, and then keep it. In this task bad news makes good copy, so that the very information that you would prefer not to be in the public domain is exactly the news that the media seek. Seldom will the media be overtly dishonest but they will want answers to the questions "what? why? and whose fault?" In the absence of reliable information from official sources, rumour and speculation, perhaps from those with an axe to grind will be used to fill the vacuum. The media will not go away if ignored, even though the public's attention span is limited, so that typically there will be a few days intense interest that will gradually subside to occasional attention as other issues rise to the boil. It is critical that the crisis management team manage those few days effectively.

A Single Voice and a Clear Message

In order to ensure that the team is free to manage the crisis and that a consistent message is put across, a single spokesperson should be identified. All briefings, press releases and media contact should be routed through that person and other team members instructed to make no comment if approached by the media.

The message that the spokesperson puts out must be factually correct and meaningful. There will need to be some internal team process for information gathering, press release drafting and authorisation by the team manager. The message that is given needs to be both simple and clear and ideally in a format that is readily usable by journalists, who are unlikely to have any particular technical knowledge, relevant to the situation. It should always provide information on:

- How the issue arose

- Why it arose
- What is being done about it

and it is important that the last point is presented as a series of positive and authoritative actions, if management is going to be perceived as being in control.

The aim should be to create a sense of confidence and good service in the minds of the media so that the natural reaction is to turn to the company for first hand, accurate information. It will be appreciated that this is diametrically opposed to the natural reaction of management in a crisis, which is defensive and protective of information, accompanied by a desire for the media to go away so that more time can be devoted to managing the crisis.

It follows that, if the aim is to manage the media and information flow, then the effectiveness of the system needs to be monitored. The results obtained will assist the team to adjust its output as necessary. This requires market research and, for a major crisis, the employment of market research consultants to give immediate feedback on performance is advisable. Their brief should be to identify the public's concerns and to confirm, or otherwise, that the message is getting across, so allowing rapid adjustment of the output by the spokesperson if necessary. Even a small company, unable to employ consultants for this task, should consider if there are not business or even personal contacts who could be asked informally to give feedback on the image that the firm is projecting. It was Shell's perception of the commercial damage that the Brent Spar dispute was having that caused the change of strategy.

When time and resources allow, a more sophisticated process that identifies different stakeholders and their individual reactions should be instituted. Amongst these we can identify:

- Employees
- Shareholders
- Institutional investors
- Customers
- Neighbours

who individually will have somewhat different concerns from each other and from the general public. Ideally each should be targeted with individual messages, tailored to meet their particular concerns and perhaps delivered by different means. So for example personal contact at board level for the institutional investor, letters to customers explaining the arrangement for continuing service and local meetings for neighbours might be appropriate. Some pre-planning can be done for this but, almost certainly the need to deploy additional professional resources on the day will mean that retaining public relations consultants is unavoidable.

Information Packs and Contacts

Pre-planning could usefully extend to having press briefing packs available as part of the emergency plan resources. These packs can contain basic information about the company and its products together with any other "good publicity" such as awards received or notable achievements. Such basic information will greatly assist the media in putting a report together, help to ensure its accuracy and minimise the load on the spokesperson. Additionally, information of a general nature about the emergency response plan might be included or more specific information about the site or process.

Finally, it is worth targeting some key individuals in the media and developing good relations with them. One might consider the editor of the local paper, radio station and the environment or business affairs reporters on the regional television companies, as being suitable. Prior and regular contact with these individuals will build up confidence on both sides and help ensure balanced and accurate reporting. Priority release of information to these individuals will mean that they can, to an extent, set the agenda and tone of media reaction.

Training

Dealing with the media is likely to be a limited skill within the crisis management team. Furthermore, the role and ability of the spokesperson are both demanding and critical. Perhaps the best investment that can be made in advance is to train the team in the skills that they will be called upon to exercise in a crisis.

The training should ensure that the team fully appreciates how the media work and that they, the team, will be expected to perform in the glare of publicity. They will also need to know the sort of message that must be projected, at whom it should be targeted and how best to do so. The spokesperson must be a confident performer at both large press briefings and one to one interviews. Training and practical exercises, making use of video and closed circuit television (CCTV), will help to identify strengths and weaknesses and to hone skills.

Reputation Protection

Many companies consider that their reputation is one of their greatest assets. It has probably been carefully nurtured over many years and may have brought the company or its products to be household names. Stakeholders come to see the reputation as being a guarantee of quality whether that be quality, of product, management or long term investment potential. Poor environmental performance can seriously undermine a reputation. One or more environmental crises could destroy it. Protection of a valuable reputation goes much further than public relations procedure practised once a crisis has broken.

However, in one sense, everything in the foregoing chapters is connected with reputation protection. The avoidance of a crisis by precaution, risk assessment and operational control, together with the management of a live crisis, can all contribute to minimising the risk to an existing reputation or help in building up a new one. Nevertheless, reputation risk management is an exercise in its own right.

Not surprisingly, the first step is to eliminate avoidable reputation risks and this requires that the risks are identified. This is directly analogous to environmental risk assessment.

Three general categories of risk can be identified. These are:

- Poor external presentation arising from poor internal communications, which lead to either inaction at critical times or worse still, conflicting action between different parts of the organisation. Better general management techniques and a procedure to review public statements before they are made should help control this risk.
- Poor environmental risk/benefit balance, which causes the company to be operating in a particular field with low commercial benefits but high reputation risks. A management operations review would identify these imbalances.
- Skeletons in the cupboard; that is known environmental risks that are simply ignored. Sometimes it may be best to proactively manage these, consulting or informing outside parties as appropriate and making public a plan to eliminate the

risk by a given date. The phasing out of CFC's announced in the annual report, is an example of what might be considered as management of a known risk.

Even if the risks cannot be eliminated, their early identification will at least buy time and allow monitoring so that a developing crisis can be identified. This gives the company the chance to prepare itself and, when appropriate, to take the initiative in protecting its reputation.

The Crisis Plan

Armed with the information provided by the risk identification and analysis, the company crisis manager can then prepare a strategic plan. The essential elements are:

- Continuous assessment of the company's position.
- Drawing together all relevant information that will assist with crisis management.
- Improvement of internal communications and ensuring that relevant personnel are fully briefed.
- Reviewing and modifying plans together with checking linkages with other plans such as the environmental emergency plan.
- Checking that systems, both procedures and physical resources, are in place and that the co-ordination with other disciplines is functioning.
- Generating a specific plan to deal with the likely environmental emergency once it can be clearly identified. This may be the site emergency plan discussed in Chapter 7 but may have to be of a more general nature to cope with a wider range of potential issues.

The reputation protection plan itself needs to be formally structured. The objective of the plan is that it should have in place:

- A trained crisis management team;
- Physical resources to enable the team to function;
- Information resources to facilitate management; and
- Agreed and tested procedures (collected and maintained in a manual,) to minimise the load on the team and allow the team to concentrate on critical issues.

Not surprisingly, these elements are almost identical to the environmental crisis management plan and clearly the two need to be closely integrated. Again it is worth making the point that the key aim of all the plans is to maintain or regain control of the situation.

The Reputation Crisis Management Team

The team needs to work as a team and be of sufficient seniority and standing to both command attention and ensure that actions are implemented. It will probably consist of:

- The Chief Executive,
- A senior manager to act as second in command,
- PR or reputation risk manager,
- Company secretary or solicitor,
- Secretarial/clerical support.

To have all these key players available all of the time and at short notice is unlikely, so understudies must be available. This team is clearly more senior than the usual environmental crisis management team. This reflects the relative roles of the two teams. The environmental one is probably site based and technically orientated. The reputation team is corporate based, has a "global" outlook and is commercially/legally orientated.

The reputation team needs to have ready access to various supporting functions and it is worth noting in passing that the environmental team could usefully also have access to equivalent functions if possible. The supporting functions are:

- PR department to act as information collector and filter,
- Managing director/operations director,
- Sales and marketing director,
- Personnel director,
- Quality assurance/customer relations director.

Physical Resources

The team needs a dedicated space from which to operate and the appropriate equipment and materials to allow it and its support to function. These differ little from the resources required by the environmental team although their scale and location may differ. Remember that the support staff may also have similar needs and that resources will be required to implement the teams' decisions. These implementation resources may be identical to those that the company uses to run its day to day business and therein can lie a problem. There is little point in the crisis team taking immediate and painful executive decisions if the actions arising from them then wait in a queue of routine work before they are implemented. The crisis management procedures should include the provision to acquire the necessary priority and company assets.

Information Resources

Similar to the environmental team, the reputation management team must have sound information on which to base its decisions. Not surprisingly, there will be an area of overlap between the two teams' requirements. Just what will be required on the day is not possible to predict in advance, so it is best to err on the side of safety and to over provide. However, an efficient retrieval system will be necessary to access the information. Obvious weaknesses such as reliance on the memory of an individual or a single computer terminal should be avoided. Ideally the system should be usable by a naive operator.

Tested Procedures

This, in effect, is the heart of emergency plan. It needs to cover the following matters:

- The crisis management philosophy of the company.
- The objectives and priorities of the management team.
- The team structure, including understudies, contact numbers, individual roles and responsibilities as well as the support that can be called on.
- The procedures to be followed, based on past experiences, and the result of simulations and exercises. At the very least these will need to cover:
  - Identifying the crisis situation
  - Assembling the management team
  - Controlling information inward and outward
  - Acquiring equipment, staff and a priority in the use of resources
  - Close-out at the end of the crisis, feedback and lessons learned.

- Management aids such as:
  - Decision trees/flow diagrams
  - Checklists of actions

- Prepared statements.

For convenience the plan may be embodied in one document that can itself become a training aid, and include a record of training undertaken. In this context it is important that the plan is immediately available to all members of the team, who must be practised in its use and have confidence in it. Such confidence will come through the testing process whereby the team deal with postulated crises either in slow time, analysing and checking the plan elements at each stage or in a real time simulation, testing the whole system and their own performance. Given the demands on the time of the senior team members, much of the development and testing would best be delegated to the understudies.

Media Management

From what has already been said it must be plain that the management of the media is a key element in reputation protection. Many of the rules which apply to the environmental teams' relations with the media apply equally to the reputation management team. However, the perspective may be more strategic and adopt a longer term view.

The principles are simple enough and set out below. It is the practice that is difficult. Firstly there is a tendency of individuals either to shy away from the media or conversely to seek unwarranted exposure. Secondly the media, while being grateful for good information, are driven by the need to find the unguarded approach or sensitive information. For example, one characteristic of a crisis is the emergence, at the behest of the media, of the embarrassing fact.

- Why was the Chairman's statement at the AGM that this had been phased out incorrect?
- Was there not an internal report produced three years ago pointing out the dangers and recommending remedial action?
- Why was Manager X moved from his post after complaining that the company was not prepared to listen to him on this matter?

This illustrates a point made by former BBC correspondant Brian Semour to a conference of emergency planners. "Remember that the media are, above all, 'people' orientated. The environment is going to be secondary to the human interest story."

Thirdly the media is not "hostile" it merely has a completely different objective to you. The protective measure are:

- Be prepared with trained staff, procedures and some established "good relations" with selected media.
- Have a strategy, but be prepared to modify the strategy in the light of circumstances.
- Respond quickly but surely in order to try and seize the initiative. Provide facilities to encourage the media to listen to you.
- Be as open and free with information as is consistent with the need to protect commercially sensitive information or avoid prejudicing your legal position.
- Provide facts rather than speculation or smokescreen. When assurances are given or action promised, make sure that they can be, and are, met.

Finally ensure tight control of information flow through the use of a competent spokesperson. Ensure also that senior staff cannot be ambushed and bounced into making impromptu statements.

For the manager who might feel that it is not his job to manage the media or even that there is something slightly underhand in doing so, remember that others will do it if you do not. BBC Scotland now admits that its reporting of the Brent Spar issue was largely in response to Greenpeace's initiatives and led to an unbalanced view being projected by the public.

Following the above rules will not prevent the crisis in the first place, nor can it guarantee that the crisis will be brought swiftly under control. They do, however, represent the best practice for minimising the potential damage done by the media to the company's reputation. Remember that the crisis represents an opportunity, if nothing else, to establish a reputation for positive and responsible crisis management, which may itself be no mean asset for the company. This point was made by Professor Kenneth Ives in his report on a pollution incident at Worcester in 1994. He noted that "Seven Trent responded with great speed and efficiency to the emergency thanks to well-prepared emergency planning and in particular to large scale exercises with simulated incidents involving staff at all levels and collaboration with health authorities. All the agencies involved worked with dedication and enthusiasm." An endorsement such as this could reasonably be reproduced in the annual report and used as a scene setter in dealings wiht the press and regulatory bodies.

Recovery

There are two separate elements in recovering from a crisis. First there are the physical actions to restore production or repair the environment. Second there are the actions necessary to restore the reputation of the firm.

Physical Recovery

Of the two this is perhaps the easiest to cope with. Where a plant or process has suffered physical damage or where there has been some failure that has led to, for example, an uncontrolled emission, the work required to rectify the situation is likely to be obvious and directly implementable either by the company's own resources or by an outside contractor. The decision as to what needs to be done may be made not by the management but imposed by the relevant regulatory authority. Management's role may be limited to that of co-ordination, particularly when damage has extended off-site. Authority for directing recovery operations will then be vested in the local authority, who may require certain works to be undertaken by the company or to undertake the work itself on a rechargeable basis.

One problem that is likely to arise in such circumstances is the difficulty of undertaking "repair" to ecological systems. It is often not only difficult to know what needs to be done but it is also difficult to be sure that what can be done is not itself damaging to the environment. In such circumstances it is best not to take hasty action, indeed no action at all may be the best strategy.

This was probably the case the Exxon Valdez when it appears that action should have been confined to containment. Unfortunately public expectation is that the "polluter pays" and pressure may well build up, aided by the media, for immediate and highly visible action that demonstrates that the company is paying a financial penalty and putting the situation to rights. The crisis management team must try and resist or redirect this pressure. In this respect, it may be advantageous if responsibility for clean-up is assumed by an outside party who can take a more considered view than the company's management when under considerable public pressure.

Recovery from long-term problems such as historical ground contamination or the removal of a product that has been marketed for some years and is now known to be environmentally damaging can be more costly, if less urgent. In these cases action may be driven by the requirement of the courts, insurers or other stakeholders, again circumscribing management's freedom of manoeuvre. Loss of control over the situation equates to loss of control over the costs and the magnitude of the crisis. Recovery must in these circumstances be on an individual basis.

Reputation Recovery

More control can be exercised over the actions taken to recover from damage to reputation although, ironically, the problem faced may be more intractable. Netherless it is possible to suggest a broad strategy for recovery which will apply in most cases.

Unlike the physical situation, reputation is longer term and there will be more time to take considered action. Pressure can, however, be expected from customers, shareholders and investors to be expeditious, particularly if share prices are suffering. Once the immediate crisis has been managed, attention can be shifted to recovery and the same management team and resources switch to the new objective.

The first requirement will be for accurate information on public perception. If consultants have been retained to assist in this respect during the crisis management phase, their brief can be extended to assess the new perception that the public has about the company following the crisis. Without consultants on hand, the public relations department might reasonably be expected to provide the same information using its contacts with the media, press reports and either selected or public interviews for example. The assessment needs to establish both the nature and extent of the change in public perception. It would be helpful to try and place the company in some sort of ranking with competitors or other major companies who can be used as benchmarks. It is worth noting that organisations such as Greenpeace produce their own Business Newsletter which gives a very clear view on its stance on various topical issues.

Armed with a clear idea of the effect that the crisis has had upon reputation,a decision can be made on the strategy for recovery . There are two basic options - separation or re-building.

Separation

This option is appropriate for situations where the damage has been substantial, so that the risk arising from being associated with crisis is too great. Such a circumstance was envisaged in the section on Reputation Protection where a poor environmental risk/benefit balance exists. The least drastic option is to consider a change of name, such that the company, a subsidiary or a product can continue without the stigma of the former crisis. Clearly, with potentially substantial costs associated with any relaunch this is not necessarily a cheap option. However, it may be preferable to the second one which is a complete break by abandoning the business, market sector or product. This strategy was adopted, for example, by the ferry company after the Zeebrugge disaster.

Rebuilding

When damage to the reputation has been less severe, then rebuilding will be the preferred strategy. Given that the reputation may be unable to withstand a second blow, priority must be given to ensuring that it does not happen again. The environmental management system or at least the crisis management system, must be reviewed and the lessons learnt applied, so as to reduce the risk of any recurrence. Once management has assured itself that it is as secure as possible from internal problems it can focus on external relationships.

Action must be tailored to suit their different stakeholders such as customers, shareholders, investors and the general public. Each are likely to have differing views of the company as a result of the crisis and hopefully the opinion survey will have provided an insight into each of them. Action can then be targeted so as to have the greatest effect. For example the regulatory authorities will expect detailed improvements in practical and managerial terms to ensure future compliance with legislation, backed up by a progressively lengthening record of achievement. The public, however, may be more receptive to money being spent on social or environmental issues that are external and perhaps peripheral to the company's activities.

British Gas maintains close links with research and advisory organisations on the environment, particularly nature conservation. It is one of the sponsors of UK Joint Nature Conservation Committee's Seabirds at Sea Project. This assists in its business planning as it can programme maritime exploration at the least damaging times. Perhaps more importantly it helps build understanding and trust between the company and regulatory or interest groups, which in turn contributes to reputation protection and enhancement.

However, it must be said that public opinion is not easily changed. We have already seen that the oil industry is perceived as a dirty one when it comes to marine pollution. Disasters such as Exxon Valdez and the Brear notwithstanding, this reputation is not really justified. Most oil companies spent substantially on the environment, even in ways unrelated to their business, such as the Shell Better Britain Campaign. This does not necessarily translate into improving either reputation or goodwill, as that company discovered with the Brent Spar. Nevertheless, none of the companies that have been used as examples in this book have chosen the separation option. All continue in their respective business sectors, some having made considerable efforts to recover their reputations.

10. *The Action Plan*

Management Commitment

Before attempting to manage potential environmental crises, it is imperative that there is a commitment by management to see the process through. Anything less will not result in a robust system, it will be a waste of money and possibly dangerous. If management subsequently believes it has adequately controlled risks, when it has in fact failed to do so, then all it has done is to create a further risk.

"The success of the system depends on commitment from all levels, especially from the highest levels of management".

So says the introduction to BS 7750 of the Environmental Management System of which crisis management is a part. Ideally, a board member or senior executive should take overall responsibility for allocating adequate personnel, time and resources to the crisis management system.

The Crisis Management Checklist

It is envisaged that in implementing a management system, the person responsible will need to produce documentation, which will have the following functions:

- Identifying risks
- Suggesting action
- Estimating resource implications

and that this will be put to senior management for endorsement. Appendix C sets out a possible structure for such a report. It is intended to give the person responsible some guidance and confidence that he will be identifying meaningful actions and responsibilities.

The template is not prescriptive and may need to be modified to take account of the particular circumstances that pertain to the individual company. It may not be necessary to complete all the sections depending on the outcome of some of the earlier ones.

It should also be borne in mind that a report based on the template will only be valid for a limited period of time, as circumstances will change. It should therefore be reviewed regularly, if infrequently.

It is unlikely that it will be possible to complete the template at the first attempt. However, it is suggested that as much as possible is completed and that the gaps in knowledge are then flagged up and actions taken to address them individually. If the company is already implementing an environmental management system, information such as that required in section 3 on legislation will be readily available from the Register. If not then it may take some time to collect such information.

Sections 1, 2 and 3 attempt to establish the base line information, what is being assessed, what the legislative controls are and who is doing the work with what purpose.

Sections 4 and 5 are the core area that is intended to identify risks and their consequences. It is unlikely that one individual will be able to complete these sections alone. The knowledge and experience of designers, site managers and technical specialists within the company should be tapped. Indeed a HAZOP or similar exercise must involve a team and a skilled chairman, who need not be the person responsible for producing the crisis management report. At 5.5 the report should be identifying priority actions based on a favourable risk/benefit ratio and outlining a strategy covering responsibilities, programme and broad resource implications.

Sections 6,7 and 8 move into the detailed actions. Section 6 is relevant if Operational Control needs to be improved or even just formalised as a set of written and auditable procedures. Section 7 covers the emergency plan or perhaps plans if, for example, a number of sites or processes are being operated. A separate appendix provides a template for the emergency plan. Section 8 considers some linkages and external interfaces. It raises the issue of reputation risks and again a separate template is provided to explore this issue further and identify an appropriate emergency plan.

The Emergency Plan Template

This is provided at Appendix D and largely speaks for itself. It may be used either in conjunction with the Crisis Management checklist or it can stand alone. In the latter case it may require additional work to complete as issues, such as a high risk plant, will not have been formally identified. Unlike the management checklist it is possible to be more detailed in this template as is dealing with a much narrower range of issues. Netherless, it has been made as general as possible and must be tailored by the manager responsible for completing it to individual site circumstances.

However, completing the template is only the first action. The plan needs to be tested, resources provided and personnel trained in its implementation. The plan arising from the template must be regarded as a draft, which then needs to pass through a review by senior management, site managers and any relevant technical specialists (production, fire, health and safety etc.) and be amended accordingly. It will then need to progress to an iterative process of testing (both theoretical and practical) and staff training, which should throw up the need for any further amendments. A final version can then be issued for use.

The Reputation Protection Plan

A template for a reputation protection plan is provided at Appendix E. Not surprisingly this bears many similarities to the crisis management template. It could also be taken as a special case of a wider reputation management plan but would, in that case, need considerable expansion. As it is necessary to make it as general as possible the author of the plan will need to modify it to suit his particular circumstances. It must also be remembered that it should be regularly reviewed.

Section 2 encourages the company to define what it is trying to protect and to differentiate between the different perceptions that the various stakeholders may have. It then goes on to identify the value of the reputation and, although it is difficult to have great confidence in this simple formula, it will help to keep matters in perspective. The level of effort and resources that are put into the plan must represent a reasonable balance between risks and benefits.

Section 3 starts by identifying the potential risks arising from environmental issues in the fields listed. Other areas of risk may be identified by individual companies involved in more specific activities. It then goes on to define the magnitude of the risk on a simple scale of five and the possible frequency. Finally some sort of priority should be established for the risks so that the plan can focus on the more critical ones.

Section 4 encourages the company to consider minimising risks before planning for crisis management. Clearly if environmental risks can be avoided by, say, plant or process modifications, procurement strategy or improved training then this should be identified and considered. Where reputation risks cannot be avoided then they can be managed by awareness and timely action intended to reduce their magnitude or slow down the development of a crisis, so allowing a greater chance of retaining the initiative and control. Once again individual companies will have their own sources of information and possible pre-emption options.

Section 5 turns to the crisis plan itself. The objective may be taken from sections 2 and 3.4, while 3.1 should identify the target stakeholders and the message that needs to be projected to each. The identification of the management team and their support follows, together with the physical resources that will be required for them to function efficiently. Since these facilities to some extent are in the nature of providing comfort to the team, the requirements may only become apparent after training and testing.

Section 6 covers the training requirements, which must include individual skills and team training in the plan execution, and the recording of training delivered. Finally Section 7 covers reporting and feedback on the plan and change control.

Again the plan will need review, testing and modification before it can be taken as complete. Reviewers should keep in mind the three essential requirements set out at the start of Chapter 7, that is they should ask themselves if the plan will:

- ensure the management team has all the facts
- allows them to gain the initiative
- enables them to control the situation.

Conclusion

The environmental crisis, when it arrives, will almost certainly be different to the one that was expected. However, given the increasing environmental legislation, stakeholder expectations and complexity of businesses, arrive it will.

Companies cannot make themselves immune to crisis, but they can prepare for it. In so doing the chance of a crisis occurring can be reduced and the chances of surviving without significant damage to themselves, or the environment, can be increased. The process of crisis management, particularly when integrated with an environmental management system, offers the potential for identifying savings to offset against the system costs. The question must be for business as it moves towards the millennium not "what can we afford to do?" but "what can we afford not to do?".

APPENDIX A

Basic References

1. Croners Environmental Management

2. Croners Environmental Policy and Procedures

3. Dealing with Disaster - HMSO

4. Dictionary of Environmental Science and Technology - Andrew Porteous
   Published by Wiley

5. Reputation Risk Management - R. Sheldon Green, F.T. Pitman

6. Environmental Management Handbook - R. Taylor, C. Hutchinson, S. Pollack and R. Tapper
   Published by Pitman for the Institute if Management and Worldwide Fund for Nature

7. Ecology - Paul Colinvaux
   Published by Wiley

8. Out of Control: Why Systems Go Wrong and How to Prevent Failures
   Published by Health and Safety Executive

9. A Guide to Risk Assessment and Risk Management for Environmental Protection - DoE from HMSO

10. Help in a Major Emergency - the SIESO Guide
    Paramount Publishing

11. a) Guidelines on Environmental Issues
    b) Guidelines on Risk Issues
    – The Engineering Council

APPENDIX B

Useful Addresses

Association of British Insurers
51 Gresham Street
London
EC2V 7HQ
Tel: 0171 600 3333

British Library
Environmental Information Service
25 Southampton Buildings
London WC2A 1AW
Tel: 0171 323 7955

Business in the Environment
(and Business in the Community)
8 Stratton Street
London
WIX 5FD
Tel: 0171 669 1600

Centre for Coastal and Marine Services
Plymouth Marine Laboratory
Prospect Place
West Hoe
Plymouth
PL1 3DH
Tel: 01752 222772

Confederation of British Industry
Centre Point
103 New Oxford Street
London
WCIA IDU
Tel: 0171 379 7400

Countryside Commission
John Dower House
Crescent Place
Cheltenham
Gloucestershire
GL50 3RA
Tel: 01242 521381

Countryside Council for Wales
Plas Penrhos
Bangor
Gwynedd
LL57 2LQ
Tel: 01248 370444

Department of the Environment (DoE)
21 Marsham Street
London
SWIP 3EB
Tel: 0171 925 5000

Department of Trade and Industry
Ashdown House
123 Victoria Street
London
SWIE 6RB
Tel: 0171 215 5000

English Heritage
23 Savile Row
London
WIX IAB
Tel: 0171 973 3000

English Nature
Northminster Road
Peterborough
PE1 1UA
Tel: 01733 340345

Environmental Auditors Association
Fen Road
East Kirby
Lincolnshire
PE23 4DB
Tel: 01790 763613

Environmental Data Services Ltd (ENDS)
Finsbury Business Centre
40 Bowling Green Lane
London
EC1R ONE
Tel: 0171 278 4745

Environmental Law Foundation
Lincoln's Inn House
42 Kingsway
London
WC2B 6EX
Tel: 0171 276 0900

Health and Safety Executive
Information Centre
Broad Lane
Sheffield
S3 7HQ
Tel: 0114 289 2345

Her Majesty's Industrial Pollution
Inspectorate for Scotland
27 Perth Street
Edinburgh
EH3 5RB
Tel: 0131 556 8400

Her Majesty's Inspectorate of Pollution
Romney House
43 Marsham Street
London
SW1P 3PY
Tel: 0171 276 0900

Her Majesty's Stationery Office (HMSO)
PO Box 276
London
SW8 5DT
Tel: 0171 873 0011

Institute of Ecology and Environmental Management
36 Kingfisher Court
Hambridge Road
Newbury
Berkshire
RG14 5SG
Tel: 01635 37715

Institute of Environmental Assessment
Gregory Croft House
Fen Road
East Kirby
Lincolnshire
PE23 4DB
Tel: 01790 763613

Institute for Environment and Health
University of Leicester
PO Box 138
Lancaster Road
Leicester
LE1 9HN
Tel: 0116 252530

Institute of Environmental Health Officers
Chadwick House
Chadwick Court
15 Hatfields
London
SE1 8DJ
Tel: 0171 823 2811

Institute of Freshwater Ecology
Windermere Laboratory
Far Sawrey
Ambleside
Cumbria
LA22 OLP
Tel: 015 394 42468

Institute of Terrestrial Ecology
Monks Wood
Huntingdon
Cambridgeshire
PE17 2LS
Tel: 01487 773381

Institute of Wastes Management
9 Saxon Court
St. Peter's Gardens
Northampton
NN1 1SX
Tel: 01604 20426

Minister of Agriculture, Fisheries and Food
3-10 Whitehall Place
London
SW1A 2HH
Tel: 0171 270 3000

Minister of Agriculture, Fisheries and Food
Public Enquiry Point
3, Whitehall Place (West)
London
SWIA 2HH
Tel: 0645 335577

National Environment Research Council
Polaris House
North Star Avenue
Swindon
SN2 IEU
Tel: 01793 411500

National Radiological Protection Board
Chilton
Didcot
Oxon
OX11 ORQ
Tel: 01235 831600

National Rivers Authority (NRA)
30-34 Albert Embankment
London
SE1 7TL
Tel: 0171 820 0101

Risk Assessment Policy Unit
Health and Safety Executive
Rose Court
2 Southwark Bridge
London
SE1 9HS
Tel: 0171 717 6403

Royal Commission for Environmental Pollution
General Office
Room Number 646
Church House
Great Smith Street
London
SWIP 3BZ
Tel: 0171 276 2080

Scottish Office
New St. Andrew's House
Edinburgh
EH3 5RB
Tel: 0131 556 8400

Royal Town Planning Institute
26 Portland Place
London
WIN 4BE
Tel: 0171 636 9107

Scottish Natural Heritage
12 Hope Terrace
Edinburgh
EH9 2AS
Tel: 0131 447 4784

Scottish Office Environment Department
27 Perth Street
Edinburgh
EH3 5RB
Tel: 0131 556 8400

Scottish River Purification Boards Association
1 South Street
Perth
PH2 8NJ
Tel: 01738 630 804

Welsh Office
Gwydyr House
Whitehall
London SW1A 2ER
Tel: 0171 270 3000

Society of Industrial Emergency Services (SIESO)
6 The Crescent
Linthorpe
Middlesborough
Cleveland TS5 6SQ
Tel: 0642 816281

Toxic Substances Division
Department of the Environment
Room A345
Romney House
43 Marsham Street
London SWIP 3PY
Tel: 0171 276 8501

APPENDIX C

CRISIS MANAGEMENT REPORT TEMPLATE

1.0 INTRODUCTION

1.1 Name of Site/Project/Process/Product
1.2 Reference/File No.
1.3 Address
1.4 Name of person completing checklist and telephone number
1.5 Name of person verifying checklist and telephone number
1.6 Copies distributed to:
1.7 Review dates 1 Signature
2
3
4
1.8 Objective of the study and its scope.

2.0 DESCRIPTION OF PROPOSAL/EXISTING SITUATION

A brief description of the salient points of the proposal or existing facility. It may be helpful to consider the matter under the headings of

Inputs
"Black Box"
Output
Waste and by-products
Consider also location and vulnerable sites and neighbours.
Where possible identify sizes, quantities, flow rates, time, frequency, distance etc.

3.0 LEGISLATION AND REGULATIONS

Is the matter covered by legislation?
What are the obligations?
What are the penalties?
What permissions, consents or licences are required? What conditions or limits?
Is it likely to be affected by emerging legislation?

4.0 UNCERTAINTY AND PRECAUTION

What are the unknowns?
Can the effects be predicted?
Are critical resources or persons likely to be affected?
Could there be substantial and unquantifiable financial risks associated with the matter?
Is the company reliant upon "luck" to avoid serious problems?
Should the project proceed in the light of the above.

5.0 RISK MANAGEMENT

5.1 What risks can be identified?

Use formal study such as HAZOP or consider each element identified in 2. above under the headings of:

* Uncontrolled discharge to air
  water
  land
* Fire/explosion
* Failure/collapse/damage in accident
* Human error
* Unforeseen effect on species, habitats or eco-systems

5.2 What magnitude?

Consider the ultimate recipient or depository for each risk in 5.1., its unreliability and whether it is statutorily permitted.
Ascribe value (see table 2).

5.3 Probability

For each risk consider likelihood of occurrence.
Ascribe value (see table 3).

5.4 Priority
Decide priorities based on product of 5.2 and 5.3 results.

5.5 So what?
Decide on which risks are significant and which are negligible. For significant risks, list possible management actions for controlling them.
Make rough estimate of cost of control and compare with risk (risk benefit analysis).
Adjust and define final actions to balance risk and benefits.
Ascribe responsibilities and programmes.
Review insurance provisions.

6.0 OPERATIONAL CONTROL

Has 5.5 above identified need for additional operational controls?
Consider procedures and work instructions.

* What is existing
* What is needed
* Who will be responsible for them.

Consider equipment, control and safety systems, their availability, use, maintenance etc.
Consider training and competency.

7.0 EMERGENCY PLAN

Use template (Appendix D) to define emergency plan. Consult, test and refine.

8.0 LIAISON

Consider reputation risks. (See Appendix E if likely to be significant).
Define means of dealing with reputation in a crisis.
Establish links with media.
Establish links with emergency planning services.

Consider if it is advisable to liase with neighbours over risk and response to emergencies.
Check relationship with Regulatory Bodies.

APPENDIX D

TEMPLATE FOR AN EMERGENCY PLAN

1.0 Immediate Response

1.1 Alarm System.

1.2 Executive instructions and flow diagram. Note this might usefully be provided to all key staff as a pocket book format.

1.3 Emergency Telephone Numbers

* Crisis Management Team
* Emergency services (Police, Fire, Ambulance)
* Regulatory authorities (NRA/HMIP (EA), EN, etc.)
* Call-off contracts (consultants and specialist firms)
* Community contacts (local authority, community groups, neighbours)

2.0 Site Information

2.1 Address and general description of works.

2.2 Location and access route, main and emergency.

2.3 Maps/plans of site location, layout of buildings, main processes and storage areas, location of emergency equipment.

2.4 Main environmental hazards - site surroundings and downstream, air, surface and ground water, physical and biological.

2.5 Copies of consents, permitted discharges etc., including plants showing locations, drainage runs, collecting tanks, outlets, stop valves, etc.

3.0 Detailed Hazards

3.1 General guidelines for dealing with emergencies - alarm, evacuation, casualties, emergency services, regulatory authorities, security, control centre and back up, information control, public relations.

3.2 Details of each individual hazard, such as:

* delivery areas
* chemical storage sites
* nature and effects of chemicals
* pipelines
* high risk plant
* accidental spillages
* map of locations
* plan of immediate action(s) at each location.

4.0 Safety & Emergency Response Equipment

4.1 Equipment - numbers, specifications and use for fire, vapours, spillage's etc.

4.2 Location - description and plan.

4.3 Operating procedures.

4.4 Communication equipment - normal communication and emergency, special telephone numbers to control information flow.

4.5 Rescue/evacuation equipment.

4.6 First aid and casualty collection.

4.7 Shelters and places of safety.

5.0 Personnel/Training

5.1 "Chain of command" for crisis management team.
5.2 Specialist staff/operators who can also be called on.
5.3 Delegated authority limits.
5.4 Competency and training requirements.
5.5 Training register.

6.0 Arrangements with Emergency Services etc.

6.1 Emergency services - roles capabilities and equipment.
6.2 Regulatory authorities - responsibilities, notification procedure of incidents.
6.3 Call-off contracts with specialists - capabilities, services, contractual arrangements.
6.4 Local community - personalities, voluntary organisations, MP and Councillors.
6.5 Dealing with the media, -who, where and how?

7.0 Inspection/Review Reports

7.1 Inspection log, - hazard sites.
7.2 Inspection log, - emergency equipment.
7.3 Plan review log.

8.0 Incident Log

8.1 Location and procedure.
8.2 Responsibility for completing.

9.0 Recovery

9.1 Activating alternative facilities.
9.2 Salvage, Repair and Replacement System.
9.3 Restart procedure

10.0 Change Control

10.1 Lessons learnt
10.2 Change control mechanism.

APPENDIX E

REPUTATION RISK MANAGEMENT PLAN

1.0 General

1.1 Author

1.2 Date of preparation

1.3 Revision status

1.4 Distribution list

2.0 What Reputation

2.1. Define reputation to be protected against environmental threats.

- Products brands
- Company image
- Central or peripheral to the company's business
- Stakeholder's perceptions
    - Shareholders
    - Regulators
    - Investors
    - Customers
    - Public etc.

2.2 Value of Reputation

- Product = buyers price minus (stock and production facilities)
- Company = buyers price minus (physical assets and expertise)
- Current cost of reputation protection measures.

3.0 Environmental Risks

3.1 Potential risks

- Legislative
- Consents/Licences
- Natural Disasters
- Plant Failure
- Product Failure
- Media Campaign
- Sabotage/Terrorism etc.

3.2 Magnitude of risk

- None
- Slight
- Moderate
- Severe
- Critical.

3.3 Frequency

- Very rare (say < once in 100 years)

- Rare (once in 100 years)
- Unlikely (once in 10 years)
- Occasional (once a year)
- Often (once a month).

3.4 Priority

Magnitude times Frequency.
Order.

4.0 Risk Minimisation

4.1 Avoidance

What environmental risks can be eliminated altogether?
How?

4.2 Risk Management

- Monitoring	- who will do it
- How	- Media
	- Industry/trade
	- Government
	- Special interest groups
- Warning signs	- What to look for
	- Who to alert
- Pre-emptive action	- Who will take it
	- What will be done
	- industry association
	- third party
	- dissociation

5.0 The Crisis Plan

5.1 Objective

5.2 Stakeholders to be targeted

5.3 Message for each covering the main risks identified

5.4 The Team

- Chairman
- Deputy
- Risk Manager
- Legal adviser
- Understudies for each of above.

5.5 Support Team

- Public affairs
- Human resources
- Sales/marketing
- Production etc. (and understudies)

5.6 Resources

- Crisis management room and equipment

- Communication facilities
  (organagrams,checklists, information facilities, decision, trees, manuals, telephones, faxes, E mail etc.)
  (who obtains information and from where?)
- Prepared materials
  (statements, briefing pack).

5.7 Spokesperson

6.0 Training

- Training requirements
- Training log

7.0 Feedback/Review

- Lessons learnt/feedback
- Review/revision process
  - who
  - programme.
- Change control mechanism.

APPENDIX F
BRENT SPAR PRESS AND MANAGEMENT BRIEFINGS

Shell UK Management Brief
North Sea abandonment - Brent Spar disposal February 1995

Greenpeace Press Release: *Brent Spar* April 1995

Shell UK Management Brief
Decommissioning of the Brent Spar: Update May 1995

Greenpeace Press Releases
Greenpeace occupies scrapped North Sea oil platform before it's dumped at sea 30 April 1995

Shell attempts to evict Greenpeace from oil platform 13 May 1995

Dumping oil platforms 'should be stopped' says EU commissioner for the environment 13 May 1995

Shell attempts to gag the press and Greenpeace – Scottish court upholds press freedom 13 May 1995

Greenpeace activists who occupied Brent Spar arrive in Shetland while legal efforts to stop the dumping continue in London 24 May 1995

UK government uses legal technicality to block Greenpeace challenge on dumping of Brent Spar 24 May 1995

Shell UK Report
Disposal of the Brent Spar: Key Facts 7 June 1995

Text of Shell UK Press Release
Brent Spar 20 June 1995

Shell UK Management Brief
Deepwater Disposal of the Brent Spar – the Shell change of plan July 1995

ISSUES
A Management Brief

North Sea abandonment - Brent Spar disposal

MAIN POINTS

- **Shell UK Exploration and Production (Shell Expro) decommissioned the Brent Spar in 1991 after 15 years' service in the Brent field as a storage and tanker loading buoy. Removal and disposal of the Spar is being planned. Meanwhile, the facility remains moored in the Brent Field.**
- **Shell Expro considered several options for disposal of the Spar. It concluded that deep water disposal was the Best Practicable Environmental Option - and the most economically efficient - after detailed examination of health, safety, economic, technical, and environmental factors by Shell UK and others. The conclusions were supported by an independent University of Aberdeen study.**
- **Following extensive consultation with the Scottish Office and other Government Departments, the Department of Trade and Industry have formally approved the deep water disposal option. The disposal is planned for summer 1995.**
- **Deep water disposal is in accordance with all UK and international laws and conventions and is supported by the Scottish Office, the Department of Trade and Industry and other Government Departments.**
- **The removal and disposal of northern North Sea platforms is going to increase, and the industry has reserved several billion pounds to meet the cost.**

The Policy Context

The North Sea is a mature petroleum province with some fields already exhausted and their facilities abandoned, and several more approaching the end of their productive lives. The numbers will increase early next century as more fields are decommissioned. Disposing safely of large deep-water platforms will involve novel technology and complex operations.

UN and IMO guidelines allow for the partial removal of offshore structures in certain circumstances, depending on water depth and the weight of the installations involved. Under UK legislation on abandonment - covered by the 1975 Petroleum and Submarine Pipelines Act and the 1987 Petroleum Act - operators must submit their preferred abandonment proposal for Government approval, together with full supporting documentation and a review of the options considered.

The petroleum taxation system requires the cost of abandonment to be borne by the field licensees, who can then offset part of the costs against tax. Where Petroleum Revenue Tax has been paid, reimbursements of previous payments could meet between 50% and 70% of the cost of abandonment. Several billion pounds have been reserved by the industry towards the cost of abandoning UK fields.

Brent Spar

The Brent Spar is a cylindrical buoy, 141 metres high (463 ft.), and weighing 14,500 tonnes. The bulk of the Spar is made up of six segmented storage tanks which were designed to hold approximately 300,000 barrels of oil and now contain sea water. The structure includes a topside with a helideck and accommodation for twenty eight. The Spar floats in around 140 metres (459 ft.) of water and is tethered to the sea bed by anchor chains, fixed to six 1,000 tonne concrete gravity blocks. Between 1976 and 1991, the Brent Spar was an oil storage facility and tanker loading buoy for the large Brent field.

Decommissioning and Disposal

The Spar was decommissioned in 1991 when it was no longer needed to handle oil from the Brent field. Without continuing maintenance, the facility would deteriorate and become a hazard. It is therefore planned to remove the Spar from the field and dispose of it in the most effective manner. Environmental impact, the health and safety of those involved, and costs must all be taken into account in evaluating disposal options.

Shell U.K. Limited

February 1995

Inventory of Brent Spar

As well as the bulk of metals which make up the body and topsides, there are other materials contained within the Spar. The storage tanks have been flushed through and the contents removed by tanker, although a very small residue of the oil once stored in the tanks may remain.

At the bottom of the tanks, there are around 100 tonnes of 'sludge' consisting of silt, which contains oil residues and small amounts of heavy metals that originated in the Brent Field reservoirs. The silt also contains low specific activity scale (LSA scale), which comprises naturally occurring radioactive salts that are a normal part of the marine environment. The silt appears as solid material on the internal surfaces of pipework and tanks. Its impact will be minimal and has no implications for health or the food chain.

Such low-level radioactivity is also present in granite rock. Brent Spar's total radioactivity is roughly the same as a cluster of granite buildings in Aberdeen.

Disposal Options

Shell Expro is committed to carrying out all its operations to the highest operational and safety standards and the Company's experience in offshore operations is second to none in the industry. The Company is also committed to carefully and responsibly evaluating disposal options for the Spar. The agreed option should be economically efficient, while complying with Government standards to protect the environment, marine safety, public health and also the health and safety of people involved in the operation.

Scrapping onshore: Detailed studies have concluded that this option poses a higher level of technical complexity and risk to health, safety and the environment.

Preparing the Spar for removal at sea, and transporting it to shore for scrapping, would involve some nineteen marine and land operations. These would include repairing some damage to storage tanks at sea, rotating the Spar to horizontal, towing to a sheltered near shore site and loading it onto a submersible cargo vessel, offloading, dismantling the structure and disposing of the waste material.

Onshore scrapping would require some 360,000 man-hours of work. The complexity and labour intensive nature of the operations means that they would be very costly and expose personnel to greater risk.

Deep water disposal: Sinking the Brent Spar at a deep water location would have minimal impact on the environment. Studies commissioned by Shell UK suggest this option is much less complicated than onshore scrapping. It involves seven, less difficult, operations requiring approximately 52,000 man hours of work.

The environmental impact of the sunken Spar will be limited and contained. The site of the sinking is likely to be in water over 2,000 metres deep and more than 240 kms from land. There is little marine life on the sea bed at such a depth. Any contaminants from the Spar would be rapidly diluted and have an insignificant impact limited to a few hundred metres from the buoy. The naturally occurring radioactive salts in the sludge and scale would have little impact on the deep sea environment.

After thorough consideration of the options, including the commissioning of an independent study, deep-water disposal was chosen as the Best Practicable Environmental Option and the most cost-effective. At a cost of more than £1 million, Shell UK is undertaking work, at the request of the DTI, survey possible deep water disposal sites and establish an environmental baseline at such sites.

SHELL VIEW

- **Abandonment plans need to be considered on their individual merits, but Government approval should be consistently based on economic efficiency, subject to compliance with realistic and reasonable environmental and marine safety standards.**
- **Each North Sea installation is unique and deep water disposal is the preferred option for the individual circumstances of the Brent Spar. Future abandonment plans will carefully consider the individual factors relevant to each case.**
- **Extensive studies and an independent report confirmed deep water disposal of the Spar as the simplest operation, posing low risk to health, safety and the environment. It would not be environmentally hazardous and would have a negligible effect on the deep sea environment.**

For additional copies of the issues brief please contact Kerensa Cobbing UKAA/14

For further information contact the Public Affairs Department (UEPA/2)

BRENT SPAR

The UK Government announced on 16 February 1995 that it had given permission for Shell to dump a huge, heavily contaminated oil installation into the sea to the west of Scotland and Ireland. Dumping operations are expected to begin in May 1995. If carried out, this grossly irresponsible act of marine pollution , on the eve of a meeting of Environment Ministers of the North Sea States, would be a clear illustration of Shell's blatant disregard for the environment and why the UK Government has the justified reputation as the Dirty Man of Europe.

WHAT IS BRENT SPAR?

The Brent Spar, owned by Shell, is an old disused oil storage installation. Under the sea surface, its cylindrical body is 109m - more than twice the height of Nelson's Column. Oil extracted from the Brent Oil Field was stored on the Spar from where it was loaded onto oil tankers. Fabricated in Norway, it was installed in 1976 and decommissioned in 1991 when it was abandoned and left to rust.

Owner Shell
Diameter 29m
Draft 109m
Height above water 28m
Total Weight 14,500t
Displacement 65,000t

The Brent Spar is located in the Brent Oil Field, in the UK Sector of the North Sea. It is 190km (118 miles) North North East of the Shetland Islands (61 03'14.7" N, 01 40'04" E).

TOXIC INVENTORY - toxic and radioactive wastes

The Brent Spar is laden with over 100 tonnes of toxic sludge and more than 30 tonnes of radioactive scale. This cocktail contains many toxic, persistent and bioaccumulative chemicals that are a serious threat to the marine environment.

MATERIAL

Amount (kg)
Waste oils 50,700
Radioactive Scale 30,000
Zinc 13,811
Copper 13,542
Cadmium 16
Lead 10
Titanium 9
Nickel 7.4
Chromium 2
Arsenic 0.3
Mercury 0.3
PCBs Trace

Source: Rudall Blanchard Associates for Shell UK 1994

PRECEDENT - tip of a toxic mountain

If the Brent Spar is dumped it will set an alarming precedent which threatens potentially huge additional toxic pollution of our seas.

There are over 400 fixed oil and gas platforms in the North Sea of which 208 are in the UK Sector. The total inventory of these structures is estimated to be:

Materials Amount (tonnes)
Steel 2,578,000
Aluminium 193,000
Stainless steel 184,000
Copper 174,000
Lead 1,800
Zinc 10,500
PCB contaminated fluids 20
Industrial gases (e.g. halons) 400
Waste oil 50,000
Radioactive scale 2,200

Source: Reddy S. Greenpeace 1995

In addition to the huge quantities of waste from obsolete oil and gas platforms, the oil and gas industry is responsible for continually discharging toxic and radioactive wastes in its day-to-day routine operation. For example, in one year an estimated 84,000 tonnes of drilling chemicals and 6,000 tonnes of production chemicals (including biocides) were dumped from North Sea oil platforms.

INTERNATIONAL OBLIGATIONS - broken promises

The dumping of the Brent Spar would be against the spirit of several international conventions that the UK is party to, including:
☐ The Geneva Convention on the Continental Shelf 1958: "any installations ...must be entirely removed"
☐ London Convention 1972: "take all practical steps to prevent the pollution of the sea by the dumping of wastes"
☐ OSPAR Convention 1992: "to prevent and eliminate pollution resulting from abandonment of offshore installations"

The dumping of the Brent Spar is expected to take place just prior to a meeting in Denmark on the 8-9 June of the Ministers responsible for the protection of the North Sea environment. Representatives from Belgium, Denmark, Germany, France, Netherlands, Norway, Sweden, Switzerland and the UK will be present. At this key time, the UK and Shell will be showing their complete disregard for the marine environment.

FLAWED DECISION - lack of detailed information

In February 1995, the UK Government decided to allow Shell to dump the Brent Spar at sea. The decision relied heavily on information supplied and paid for by Shell itself. This information has been heavily criticised by off- shore industry experts who state that the technical case for dumping is unproven . Among the concerns are:
☐ no detailed survey or structural analysis has been carried out,the toxic inventory is not accurately known and is likely to be a gross underestimate
☐ cost estimates for the various options are unlikely to be accurate;disposal plans have been formulated without adequate consultation with marine engineers
☐ it is technically feasible and environmentally beneficial to remove off- shore installations to land

In short, the decision to dump the Brent Spar has been based on limited information.

FEASIBLE ALTERNATIVE - complete removal to shore

The Brent Spar was originally constructed vertically in a deep water site in Norway. Of the numerous alternatives to dumping, the most obvious is to reverse this process, i.e. tow to a deep water site and dismantle from top to bottom; thus allowing responsible management of the resultant wastes.

This technically feasible option has a number of benefits:
☐ it causes the least damage to the environment
☐ it will bring much needed employment to traditional marine construction industry areas
☐ it is in line with international obligations, international bans on the dumping of radioactive and industrial wastes and promises made to remove such installations
☐ an enormous local recycling potential of high grade steel etc.

In the US, Federal Law insists that off-shore platforms in the Gulf of Mexico are completely removed. Since 1987, 914 platforms have been removed including at least 20 oil installations owned by Shell. It appears that Shell has more concern for US waters than it does for European seas. In contrast to the UK, the Norwegian Government has stated its preferred option to remove a similar structure to the Brent Spar, called N.E. Frigg, for on-shore dismantling and recycling.

ECONOMICS - costing the sea

The lack of detailed studies carried out by Shell means that cost estimates for the various options are in doubt. Experience gained by the off-shore engineering contractors would indicate that the reality of removal is less costly than originally budgeted. In spite of this, Shell estimates that complete removal of the Brent Spar would cost œ46 million. This represents only 3.5% of the œ1.3 billion that Shell has been spending on refurbishing its other installations in the Brent Field since 1993. The value of oil and gas sales from the North Sea in 1993 was œ13.8 billion. In 1991, Shell was the only oil company to have assets in excess of $100 billion.

OUR SEAS IN DANGER

Marine ecosystems, notably of the North Sea, are in serious decline. Chemical pollution from industrial discharges, sewage, radioactive discharges, pesticide run-off, atmospheric pollution are all continuing to enter our seas.

Given these concerns it is particularly alarming that the UK Government should be allowing Shell to dump the Brent Spar. The pollution from the off- shore Oil and Gas Industry remains poorly regulated by North Sea governments. The dumping of the heavily contaminated Brent Spar clearly shows a complete disregard for the health of the marine environment by the UK Government and the Oil and Gas Industry.

GREENPEACE DEMANDS

☐ The UK Government should immediately reverse its decision to permit Shell to dump the Spar
☐ Shell should immediately reverse its decision to dump the Brent Spar. It should be completely removed and its wastes managed responsibly on land
☐ Further urgent measures should be agreed by the North Sea governments at the forthcoming Ministers Meeting in June to ban the dumping of off-shore oil and gas platforms and prevent the routine operational hazardous discharges from this highly polluting industry.

Greenpeace April 1995

ISSUES
A Management Brief

Decommissioning of the Brent Spar: Update

MAIN POINTS

- **In February this year the Government outlined its policy for the decommissioning of redundant offshore petroleum installations. It also approved a proposal from Shell UK Exploration and Production to dispose of the Brent Spar by sinking it at a predetermined location in the North Atlantic, at a depth of over 6,000 feet and a distance of some 150 miles from the West Coast of Scotland. Subsequently the Scottish Office has issued licences which complete the process of formal approvals.**

- **The Spar, a floating storage and tanker loading buoy anchored in the Brent Field, was decommissioned in 1991, and much of the equipment and remaining sources of pollution have already been removed. Before it is towed away from the field, three more weeks of offshore work are planned to remove, as far as is practicable, the regulated and other movable materials which still remain on board. A full inventory, which forms part of the disposal submission to the Government, has been established for this purpose.**

- **The Government endorsed this plan after Shell committed more than three years to painstaking analysis of the options, the completion of fifteen separate studies on their environmental and engineering aspects, taking independent surveys and advice into account, and the careful weighing of all the environmental, safety, public health and economic considerations. Every aspect of UK law and regulation has been followed to the letter and to the spirit. The Government is entirely satisfied that it is also complying with the relevant aspects of international law and conventions.**

- **Shell UK has also spent more than £1 million on surveying alternative deepwater sites for the sinking. Deepwater disposal of the Brent Spar does not create a precedent. In accordance with Government policy, future decommissioning plans will be considered on a case-by-case basis in relation to environmental, safety, public health and economic considerations. Each installation is unique and the preferred options are likely to differ. Indeed Shell Expro's current view is that very few structures will be disposed of through sinking in an Atlantic site.**

Background

The background is given in the issues brief 'North Sea Abandonment - Brent Spar disposal' dated February 1995. A review of the Spar's operations in 1991 concluded that, after 15 years of operation, there was a case for decommissioning the installation and ultimately abandoning it in accordance with Government policy. The facility was therefore decommissioned in September 1991.

Environmental Considerations

Since then most of the fixtures and other scrap have been removed, and the transformers have been drained of fluid which contained PCBs. The empty storage tanks have been flushed through and the contents removed by tanker. All practicable steps will be taken to remove and safely dispose of the remaining sources of contamination before the deep sea disposal; the contents of the inventory mentioned above will be worked through systematically. Some three weeks of work is planned to complete this. Since decommissioning, the facility has been regularly checked and maintained to ensure compliance with the applicable marine safety standards.

The irreducible sources of possible contamination sunk with the Spar will consist of the paints and sacrificial anodes (corrosion preventors made of zinc) on the structure itself; up to 100 (not 300 as has been alleged) tonnes of sludge, consisting of 90% sand and 10% oil residues containing very small quantities of heavy metals; and 30 tonnes of solid deposits of low level radioactive salts in the tanks and pipework. These salts occur naturally in geological structures and can deposit, from produced water associated with the oil, a scale on the inside of pipework and tanks. Altogether these represent such low levels of radioactivity that they fall many orders of magnitude within international standards. The level of radioactivity, for example, will be no more than would emanate from a group of granite buildings in a city such as Aberdeen. Furthermore, the radioactive scale would only be dangerous to humans if allowed to dry out and form dust which could be breathed into the lungs in

Shell U.K. Limited

May 1995

large quantities. This was one of the factors which weighed against onshore scrapping of the facility.

The unequivocal conclusion from the independent analyses of the options and of the environmental impact of the deep sea disposal is that the environmental impact will be both negligible and localised, any contamination staying within a few hundred metres of the seabed disposal site.

Contrary to some reports, it is established practice in some countries to permit the option of sinking or toppling of redundant offshore petroleum installations. In the United States, for example, the Government has since the mid-eighties encouraged oil companies to topple some 95 structures in the Gulf of Mexico to create artificial reefs, on which marine life has proliferated, specifically to the benefit of fishermen. Nonetheless, in the case of the Brent Spar, the extreme depth at which it will lie means that there is likely to be very little fish life.

Formal consultations were carried out on the deep sea disposal option with a number of organisations in 1994. These included The Scottish Fishermen's Federation, Scottish Natural Heritage and the Joint Nature Conservation Committee. All those consulted supported the option. Following the consultation period, Shell Expro has made the Best Practicable Environmental Option (BPEO) assessment and Environmental Impact Assessment, both of which supported the formal submission to Government, available to the public, the media and other interested organisations (including Greenpeace).

The Options

These are described in detail in the previous issues brief. In essence, the onshore option would involve 19 separate, and technically complex, land and marine operations. These would include repairing some damaged storage tanks, towing the buoy into environmentally sensitive coastal waters, and there rotating it to a horizontal position for dismantling and disposal. The 360,000 man-hours needed would increase the risk of an accident or an environmentally damaging incident by an order of six times. Most of the waste materials would end up in landfill. This option would cost four times as much.

The option of sinking it to the deep seabed of the Atlantic Ocean would involve seven, less difficult, operations requiring some 52,000 man-hours. Personnel would be at less risk and on balance the impact on the environmε would be negligible in both cases. The environmental effect of the sinking itself will also be negligible.

Leaving the buoy where it is was not considered to be a viable option from any point of view - environmental, marine safety or economic. In any case, it is a licence condition that redundant offshore petroleum installations must be abandoned in accordance with a plan approved by the Government following consultation with interested parties.

SHELL VIEW

- **The disposal of the Brent Spar by sinking it at a remote deep water Atlantic location is the best practicable option taking into account all the environmental, safety, public health and economic considerations. Before sinking, the structure will be stripped of almost everything on board and the residual sources of contamination will be minimised.**
- **The Government's authorisation of this method of disposal does not set a precedent and is likely to be used in very few cases for other offshore installations. Modules recently removed from one of the Brent platforms were, for example, brought ashore and dismantled.**
- **The disposal has been authorised by the Government as being in accordance with all the relevant national and international laws, agreements, regulations and conventions.**
- **In-house and independent assessments and studies have concluded that the impact on the marine environment will be negligible. Fishing and environmental organisations consulted agreed with our analysis.**

For additional copies of the issues brief please contact Carolyn Smith UKAA/16

For further information contact UEPA/2

GREENPEACE OCCUPIES SCRAPPED NORTH SEA OIL PLATFORM BEFORE IT'S DUMPED AT SEA

NORTH SEA, SUNDAY, 30 APRIL 1995--Greenpeace today scaled and took up residence on an old North Sea oil platform to stop its owner, Shell, from dumping the rusting hulk and its highly toxic contents to the sea bed.

Four climbers used ropes and winches to scale the Brent Spar, which is the first of the 400 North Sea oil platforms to be dumped at sea. The climbers have food and supplies for what is expected to be a long occupation. The Moby Dick is standing by as a safety vessel. The activists first climbed the steel ladders on the installation, then scaled the outer part of the rig, 28 metres high above the waterline.

After Moby Dick captain Pelle Pettersson notified the five other rig support vessels in the area, stating that Greenpeace was protesting the dumping of the Brent Spar, a Shell standby vessel then sailed very close to the Moby Dick and the inflatables, harrassing them.

The Brent Spar contains over 100 tonnes of toxic sludge --including oil, arsenic, cadmium, PCBs and lead -- including more than 30 tonnes of radioactive waste left over from oil drilling and storage operations on the Brent Oil Field.

The UK Government's decision to allow Shell to dump the Brent Spar was based on information supplied and paid for by Shell.

Ironically, the planned dump comes just one month before North Sea environment ministers meet in Denmark in June to discuss measures to eliminate the discharge of hazardous substances from all sources into the North Sea and the marine environment.

"The decision is short-sighted and the latest example of governments allowing industry to treat the seas as a toxic dump," said Greenpeace's Tim Birch onboard the Moby Dick.

"Greenpeace will remain on Brent Spar until the UK Government or Shell comes to their senses and revokes the decision to dump it," Birch said.

A Greenpeace report released today, "No Grounds for Dumping: The decommissioning and abandonment of off-shore oil and gas platforms", analyses the decommissioning options available to the UK Government. The report concludes that total removal is not only the best environmental option but also the most cost-effective, feasible and job-saving.

"If the UK Government does not reverse its disastrous decision, dumping the toxic- laden Brent Spar at sea could set a domino effect in motion as more of the 400 North Sea oil platforms are abandoned," said Birch.

Marine ecosystems, especially in the North Sea, are still under serious pressure from chemical and radioactive discharges, sewage pollution, pesticide run-off and atmospheric pollution, among other sources.

"Dumping oil platforms laden with toxic cocktails will only add to the list of contaminants and proves yet again that the UK Government and the oil and gas industries have total disregard for the health of our seas," said Birch.

ends

For more information and/or a copy of the Greenpeace report: Tim Birch or Desley Mather onboard the

Date: Sat 13 May. Time: 03.32 GMT

SHELL ATTEMPTS TO EVICT GREENPEACE FROM OIL PLATFORM

Greenpeace stands firm on North Sea protection

LONDON, 13 May 1995--Greenpeace last night vowed to continue its North Sea campaign to fight the dumping of the Shell Brent Spar oil platform, despite a legal attack from the oil company.

At 9.30pm last night, a Shell helicopter landed on the Brent Spar with two Sherrifs officers from Aberdeen carrying a legal injunction and wearing Shell survival suits. As the helicopter hovered over the platform, a 12 page legal document was faxed to the 13 Greenpeace activists who have occupied the platform for the past 13 days. Greenpeace lawyers are currently studying the document.

"Whatever strong-arm legal tactics Shell choose to use, Greenpeace will continue to oppose the dumping of the oil industry's toxic garbage in the North Sea," said Greenpeace spokesperson Chris Rose.

"Shell's attempts to use the courts to enforce its disastrous plan to pollute the seas demonstrates its contempt for public concern about its operations, fishermen's livelihoods and for the health of the North Sea," Rose said.

"Shell is chasing cash at the expense of the North Sea marine environment -- yesterday it announced a 35% increase in profits to 1280 million in just the first three months of the year. To dump the Brent Spar as a cheapskate alternative to responsible decommissioning with decontamination onshore is nothing short of obscene," said Greenpeace.

The Brent Spar contains over 100 tonnes of toxic sludge -- including oil, arsenic, cadmium, PCBs and lead -- and more than 30 tonnes of radioactive waste left over from oil drilling and storage operations on the Brent Oil Field. It is the first of the 400 North Sea oil platforms to be dumped at sea.

For more information, please contact Chris Rose on 0860 355271

For photographs of the Sherrif's arrival onto the platform, please contact Reuters, AP and PA London or Greenpeace ++44 71 833 0600

Date: Sat 13 May. Time: 16.24 GMT

FOR IMMEDIATE RELEASE - 13.5.95

DUMPING OIL PLATFORMS "SHOULD BE STOPPED" SAYS EU COMMISSIONER FOR THE ENVIRONMENT

- International support for Greenpeace occupation of Brent Spar Grows

International support for Greenpeace action to prevent the sea dumping of toxic laden North Sea oil platforms continues to grow.

Last night Ritt Bjerregaard, EU Commissioner for the environment, appeared on Danish TV. Also last night Shell launched a legal attack on the occupiers of the Brent Spar in an attempt to force them to leave. Greenpeace is seeking legal advice.

Quotes translated from Danish TV:

Ritt Bjerregaard, EU Commissioner for the environment:

"Actually, I think most countries in the EU thinks this is dirty and that it should be stopped. If it is banned in the US they try other places and unfortunately there is both countries and enterprises in this situation who choose the cheapest option and we have to make sure they do not have that opportunity... is the simple thing to do and in the short term you avoid the problems with some firms and therefore they choose that. Therefore it is good that Greenpeace is around to ensure these things do not go on secretly."

Svend Auken, Danish Minister for Environment and Energy:

"We think there will be many hazardous substances which will escape into the enviornment this way, and in any case the sea should not be used as a dustbin....There are many problems in the North Sea. Many of the platforms out there are about to reach the time when they will have to be scrapped. This has to be done in a safe manner and the only way this can be done is on land. We cannot accept just toppling them at sea... The fact that the host of the Ministers' Conference raises this matter, that we support Greenpeace and others, and that we are supported by other countries, that will have to make an impact on those considering dumping, I think..... ."

Peter Sand Mortensen, Chair of the Fishermens Sector, International Transport Federation, ITF.

"It is quite simply a catastrophy. For the environment, for the fishermen , for all of us....if a permit is given now, what is to be expected for the 400+ other installations out here. Economic factors are working in favour of just leaving everything out there once profits have been gained....this involves one of the largest, Shell."

For Further Information contact Sue Cooper: 0171 359 9980 or Chris Rose 01595 694 099.

For Immediate Release - 13th May 1995

Date: Sat 14 May. Time: 17.11 GMT

SHELL ATTEMPT TO GAG THE PRESS AND GREENPEACE - SCOTTISH COURT UPHOLDS PRESS FREEDOM

In a legal attempt to end Greenpeace's 14 day old occupation of their North Sea oil installation, the Brent Spar, Shell asked the Scottish Courts to issue an order gagging the press.

Shell requested "an order in terms of the Contempt of Court Act 1981, Section 4(2) postponing the publication of any report relating to these proceedings for such period as the Court shall consider appropriate." Lord Kirkwood, the Scottish Judge, did not grant their request.

Shell did manage to obtain an interim interdict attempting to prevent the Greenpeace vessel Moby Dick from resupplying the Brent Spar occupiers with food and other essentials.

"Shell's tactics are secrecy and starvation. They are clearly trying to force the people off the Brent Spar," said Chris Rose, Greenpeace spokesperson.

"Shell are so ashamed of what they are doing that they sought to do it behind a veil of secrecy. Lord Kirkwood, in denying Shell's request, has recognised that the dumping of toxic oil platforms at sea is a matter of great public interest. He has decided that if Shell are to do any dirty business it must be done in public," said Rose.

The international group of activists on board Brent Spar are awaiting legal advice, together with copies of the court orders written in their respective languages.

Greenpeace is determined to draw attention to the dumping of toxic-laden platforms at sea.

For further information: Sue Cooper 0171 359 9980 or Chris Rose - 01595 694 099. For legal comment contact one of the above numbers.

- ☐ Return to Brent Spar occupation home page
- ☐ View previous press release
- ☐ View following press release

Date: Wednesday 24th May. FOR IMMEDIATE RELEASE

GREENPEACE ACTIVISTS WHO OCCUPIED BRENT SPAR ARRIVE IN SHETLAND WHILE LEGAL EFFORTS TO STOP THE DUMPING CONTINUE IN LONDON

LONDON, 24 May, 1995, 1:30 BST -- (GP) The remaining Greenpeace activists who occupied the Brent Spar in protest of plans to dump the toxic oil platform at sea have just arrived in Shetland.

The first group of eight activists and six journalists were taken by Shell helicopter yesterday from the Stadive to Shetland. None of those returned have been charged. The remaining Greenpeace activists on the Stadive refused to leave until all were safely off the Brent Spar.

Shell told reporters that Jon Castle will be taken to Aberdeen this evening from Shetland and will then be released.

Greenpeace is currently in court proceedings in the High Court in London to attempt to challenge the government's decision to allow the dumping of the highly toxic and radioactive oil platform at sea.

"Although our occupation of the rig has ended, the fight to stop the dumping of the Brent Spar oil platform is not over yet," said Jan Rispens, one of Greenpeace's Brent Spar activists. "We continue to work to stop the dumping in the courts and at the North Sea Conference."

Criticism against Shell and the UK government's plan to dump the Brent Spar continues to grow with Germany adding its voice to those of the EU Environment Commissioner, European Parliament (EP), Denmark, Belgium, Iceland and even the UK's own Labour Party. The issue is likely to be hotly debated at the upcoming meeting of the North Sea Conference in Denmark (June 8-9).

"The occupation of the Spar has again highlighted Shell and the UK government's outrageous attitude towards our seas," Rispens said. "They seem determined to take the cheapest option rather than the safer alternative known to them."

Contact: Mary Morrison, Greenpeace Shetland: ++44(0)-1595-694-099
Simon Reddy, Greenpeace Aberdeen: ++44-0374-695-814
Blair Palese, Greenpeace Communications: ++44(0171)-833-0600

Immediate Release

UK GOVERNMENT USES LEGAL TECHNICALITY TO BLOCK GREENPEACE CHALLENGE ON DUMPING OF BRENT SPAR

LONDON, 24 May 1995, 4 pm -- (GP) The High Court quashed any possibility of a Greenpeace challenge to the UK Government's decision to allow the dumping at sea of the Brent Spar oil platform in a British court today.

The UK Government and Shell had argued that the case should be heard in the Scottish courts. After nearly a full day in court, the judge ruled that the English courts had no jurisdiction over the case despite the fact that he agreed Greenpeace would have no standing in the Scottish courts. The judge suggested the case be taken to Scotland anyway.

"This decision, based on a legal technicality, means that no British court can hear Greenpeace challenge the Government. In effect it is a decision by the UK Government to allow Shell to dump toxic chemicals and radioactive waste at sea," said Madeleine Cobbing of Greenpeace. "If the Government could justify what it is doing it could defend itself in court. It is obviously too frightened to argue the case with

Greenpeace."

Criticism against Shell and the UK Government's plan to dump the Brent Spar continues to grow with Germany adding its voice to those of the EU Environment Commissioner, European Parliament (EP), Denmark, Belgium, Iceland and even the UK's own Labour Party. The issue is likely to be hotly debated at the upcoming meeting of the North Sea Conference in Denmark (June 8-9).

"The UK Government can still stop the dumping of the Brent Spar if it has the courage to take the right decision. We call on them to respond to the International condemnation from other Governments, and cancel the dumping of the Brent Spar immediately."

As the last of the Brent Spar occupiers returned to the mainland, Greenpeace pledged to continue its campaign against the dumping of the Brent Spar and of all hazardous substances into the sea.

Calls from other North Sea States to eliminate the discharge of hazardous substances into the sea are still being resisted by the UK Government. A Greenpeace report* released tomorrow reveals that the UK is still the worst polluter of the North Sea.

Contact:
Sue Cooper, Greenpeace UK: (44) 0171 354 5100
Blair Palese, Greenpeace Communications: (44) 0171-833-0600

*"The North Sea - A Dirty Story" -- for copies, contact the numbers above.

- ☐ Return to Brent Spar occupation home page
- ☐ View previous press release
- ☐ View following press release

Disposal of the Brent Spar – Key Facts

Claims being made by Greenpeace in their campaign against the Government-approved disposal plan for the Brent Spar oil storage and loading buoy are alarmist and incorrect. The following facts cover the concerns raised so far:

- UK policy and regulation for the disposal of redundant offshore oil and gas installations comply fully with the highest international standards.
- The Spar disposal plan, entailing painstaking analysis of the options; responsible balancing of all environmental, safety, health and economic factors; and extensive consultations with interested parties including fishermen and environmentalists, reflects best international oil industry practice.
- Future disposals will be decided, case-by-case, in the same rigorous way and onshore recovery is generally foreseen - and already practised.
- The exceptional choice of deepwater disposal for Spar was made because numerous studies have shown that while there would be no environmental benefit in onshore disposal, safety and occupational health risks would be six times greater.
- On safety grounds, there are risks that having been damaged in operation and possibly overstressed during installation, the Spar might break up while undergoing the technically difficult task of upending for onshore disposal. On occupational health grounds, the very low radioactivity scale in the Spar could be dangerous if inhaled as dust during scrapping - left as it is, it poses no risks.
- The carefully surveyed and selected disposal site is 150 miles out in the Atlantic (not the North Sea) in a water depth of around 7800 feet (2375 metres or about 1.5 miles).
- Contamination to the sea bed is independently assessed to be negligible, very localised, and inaccessible to the food chain because at this depth marine life feeds off organic debris sinking from above. To the last light bulb, all sources of contamination which can be removed safely are being recovered before disposal of the Spar.
- The 100 tonnes of sludge (highest estimate) consists of 90 per cent sand, and 10 per cent stable, heavy oil residues - not dissimilar from the bitumen on roads. The tiny proportion of heavy metals in the sludge is not significantly different from that which would be found naturally in the same weight of plankton or marine sediments.
- The very low level radioactivity from scale inside the Spar (highest estimate 30 tonnes) is equivalent to that from a group of granite houses in Aberdeen; 2 kg of scale equates to the radiation from a domestic smoke detector.
- The benefits of recycling the metal from Spar have been evaluated. The analysis shows that there is no significant benefit in terms of overall energy and resource savings.
- Consultations with fishermen and environmentalists before the deepwater disposal was approved indicated support in this case. None of the European governments notified by HMG under the Oslo Convention queried the plan before the Greenpeace action.
- Removal of redundant offshore installations is not a universal regulatory requirement. The US Government, for example, supports the careful placing of redundant structures, in some cases, to create artificial reefs which are attractive to marine life and benefit fishermen in a sustainably responsible way.

More detailed background is available from the Issues Briefs of February and May 1995; from Chris Fay's letter to Greenpeace International; and from the Best Practical Environmental Option analysis of December 1994. Please contact UKAA/16 if required.

UKAM June 7th, 1995

Text of Shell UK Press Release issued at 17:49, 20 June 1995:

Shell UK Limited,
Shell-Mex House,
Strand,
LONDON WC2R 0DX

For immediate release: June 20 1995

BRENT SPAR

Shell U.K. Limtited still believes that deep water disposal of the Brent Spar is the best practicable environmental option, which was supported by independent studies. Shell UK obtained the necessary permit for disposal for the UK authorities, which have handled every aspect of the approval process in accordance with established national and international policies and standards.

As the disposal involved the Atlantic deep water, other governments have taken an interest and voiced strong objections. Notwithstanding the efforts to convince these governments of the validity of the approach, most of them remain strongly against deep water disposal.

The European companies of the Royal Dutch/Shell Group find themselves in an intenable position and feel that it is not possible to continue without wider support from the governments participating in the Oslo-Paris Convention.

Shell UK has decided to abandon deepwater disposal and seek from the UK authoities a licence for onshore disposal. This application for onshore disposal wil include a further review of methods to minimise the risks involved.

The way forward will require the co-operation of all concerned, in particular the support of other governments in securing the availability of a deep water safe anchorage for Brent Spar in the immediate future.

ENQUIRIES

Shell UK Media Relations 0171-257-3152/3333/3889/3100

☐ Return to Brent Spar occupation home page

ISSUES
A Management Brief

Deepwater disposal of the Brent Spar – the Shell change of plan

The controversy surrounding plans for deep water disposal of the Brent Spar in the Atlantic highlighted the difficulties facing those responsible for seeking environmental solutions. Careful analysis of environmental, safety, health and economic considerations had convinced independent experts, Shell UK and the British Government that this was the best option. An emotive Greenpeace campaign gave it a symbolic significance beyond any rational, scientific calculation of its impact. Political and public pressure on Shell companies on the Continent, and the growing safety threat, forced Shell UK to abort its plans.

This has been a considerable setback for Shell UK, an organisation committed to environmental improvement, and to the goal of sustainable development pursued through sound science, reason and the careful balancing of environmental, safety, health and economic considerations. We must learn from what has happened. But there are also wider lessons for business, regulators, environmentalists and society as a whole. Environmental solutions must gain the support of people's hearts as well as minds, but a rational scientific approach to the environment is essential.

On 20 June 1995, within days of the intended deepwater disposal of the Brent Spar oil storage and loading buoy in the Atlantic Ocean, Shell UK aborted the operation because the Shell position as a major European enterprise had become untenable.

After three years of painstaking analysis, planning, and extensive external consultation by Shell UK, the deepwater disposal plan had been independently assessed as the Best Practicable Environmental Option. Under the case-by-case approach which is consistent with the international Oslo and Paris Conventions governing the protection of the marine environment, all considerations – environmental, safety, occupational health, and economic – favoured deepwater disposal.

In a well orchestrated campaign, Greenpeace had used misinformation, together with visually dramatic stunts and imagery, to give the deepwater disposal of the Brent Spar a symbolic significance out of all proportion to its environmental impact. For example, the impact of sinking the Spar at 1.5 miles depth in the Atlantic Ocean would have been minuscule in comparison with the river-borne pollution inflicted daily on the much more sensitive and shallower North Sea.

In consequence, Shell companies were faced with increasingly intense public criticism, mostly in Continental northern Europe. Many politicians and ministers were openly hostile and several called for consumer boycotts. There was violence against Shell service stations, accompanied by threats to Shell staff.

The difficult decision to abandon deep-water disposal because the Shell position as a major European enterprise had become untenable was not driven by financial expediency. It was based on two key considerations.

Firstly, the disposal plan was based on scrupulous compliance with established national and international regulatory principles and standards. Yet most of the Continental northern European governments which are parties to the Oslo and Paris Conventions, and had originally raised no objection to the plan, were now openly opposing it – not on its technical merits, but because of its symbolic significance in the light of the Greenpeace campaign. This became clear in an urgent round of further Shell consultations with European governments in mid-June.

Secondly, an important consideration which had underpinned the choice of deepwater disposal was the much greater risk to the health and safety of employees who would have carried out the alternative onshore disposal plan. Yet actions on the Continent, and by Greenpeace activists attempting to interfere with the Spar deepwater disposal, now represented a serious safety threat.

The decision to abandon deepwater disposal operations on June 20 was immediately conveyed personally to Tim Eggar, the British Industry and Energy Minister. John Jennings, the Chairman of The "Shell" Transport and Trading Co, plc, wrote to the British Prime Minister John Major apologising for the position in which the outcome of this complex issue had placed him and his Ministerial colleagues.

Shell U.K. Limited

July 10 1995

On the eventual disposal of the Brent Spar, the UK Government has made it clear that it expects Shell UK to base any alternative disposal plan put forward for approval and licensing on the same rigorous analysis and planning, incorporating any new ideas, as went into the original deepwater disposal plan – assessed as the Best Practicable Environmental Option. This is likely to take a good many months, during which Shell UK will seek to hold even wider external consultations in order to ensure that formal licences are backed by the wider support of stakeholders generally.

In this connection, Greenpeace has written to Shell UK, offering their involvement in the development of any alternative disposal plan. Greenpeace has been advised that Shell UK is prepared to consider including them in the list of external bodies to be consulted at the appropriate stage in the process – in exactly the same way as other bodies and environmental groups.

Numerous offers have been received from contractors seeking contracts for work in connection with the future disposal plan. They have been advised that a great deal more analysis and planning is required before Shell UK Exploration and Production (Shell Expro) is in a position to involve contractors. Thereafter, their proposals will be considered on merit in accordance with Shell Expro's normal arrangements.

Approval has now been granted by the Norwegian authorities for a temporary safe anchorage in Erfjord, in Rogaland County (east of Stavanger). The Spar is expected to arrive at Erfjord by mid-July.

Beyond the immediate issue of the disposal of the Brent Spar lies the wider issue of the international regulatory principles and standards for the removal and disposal of redundant offshore petroleum installations. The established policy, embodied in International Maritime Organisation guidelines and standards, is that any installation in shallower waters of less than 75 metres must be totally removed. For installations in deeper waters, the principle is that each case for disposal must be considered on its merits, with regard to environmental, safety, occupational health and economic considerations – as in the case of the Brent Spar. Some heavier structures in deeper waters might be partially removed, but any remains must be at least 55 metres below the surface.

There are currently 219 installations in UK offshore waters. The great majority of these are in shallower water and would be disposed of onshore. Some 50 deepwater installations would be assessed in the same way as the Brent Spar, with the full range of options evaluated. In all decommissioning plans, operators must demonstrate under the Offshore Installations Safety Case Regulations how the safety risks involved will be reduced to as low a level as is reasonably practicable.

At the meeting of the Oslo and Paris Commissions in Brussels on June 26-30, 11 states supported a moratorium on the disposal at sea of decomissioned offshore installations. This was opposed by the governments of Britain and Norway, the only two countries which have offshore petroleum installations in water of more than 75 metres.

Shell UK believes that the wider issues surrounding the abandonment of the Brent Spar disposal operation present far reaching questions, not only for industry and its regulators, but for society as a whole. What role should emotive and partial campaigns by 'single issue' campaigners, focusing on just one aspect of complex problems, play in the democratic process? What responsibility do they have for ensuring that the best practicable solutions are chosen? Can they contribute to the essential process of reasoned discussion, meticulous evaluation and careful consultation? And are they willing to heed scientific judgements rather than irrational fears?

The notion that environmental choices can be made without regard to costs is unsustainables. Society faces many challenges and if money and resources are spent disproportionately on one problem it will be more difficult to tackle others. Employment, standards of living, social provisions and environmental progress all depend on economic health. For Western Europe, declining international economic competitiveness is a vital concern.

However, there are also positive and potentially constructive aspects to the Brent Spar events. Businesses face the challenge of finding new means of including in their planning and consultations not only the views and rational arguments of all concerned, but also deep-seated emotions and subconscious instincts which some projects, however soundly based, may evoke – of gaining the popular support of hearts as well as minds. The events may also serve to focus attention on the complexity of environmental questions, on the need for sound science as a basis for decisions, and on the consistent regulatory framework which businesses need in order to invest and to sustain employment.

For additional copies of the issues brief please contact Carolyn Smith UKAA/16